Fascinating Life Sciences

This interdisciplinary series brings together the most essential and captivating topics in the life sciences. They range from the plant sciences to zoology, from the microbiome to macrobiome, and from basic biology to biotechnology. The series not only highlights fascinating research; it also discusses major challenges associated with the life sciences and related disciplines and outlines future research directions. Individual volumes provide in-depth information, are richly illustrated with photographs, illustrations, and maps, and feature suggestions for further reading or glossaries where appropriate.

Interested researchers in all areas of the life sciences, as well as biology enthusiasts, will find the series' interdisciplinary focus and highly readable volumes especially appealing.

More information about this series at http://www.springer.com/series/15408

Harry J. Flint

Why Gut Microbes Matter

Understanding Our Microbiome

 Springer

Harry J. Flint
Rowett Institute of Nutrition Health
University of Aberdeen
Aberdeen, UK

ISSN 2509-6745   ISSN 2509-6753   (electronic)
Fascinating Life Sciences
ISBN 978-3-030-43248-5   ISBN 978-3-030-43246-1   (eBook)
https://doi.org/10.1007/978-3-030-43246-1

This Springer imprint is published by the registered company Springer Nature Switzerland AG.
The registered company address is: Gewerbestrasse 11, 6330 Cham, Switzerland

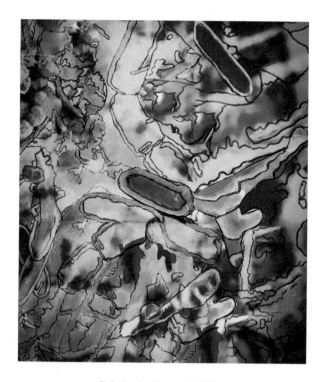

Painting by Rowan M. Flint

In memory of my father Harry Edgar Flint (1924–1988).

Preface

Until recently, the micro-organisms that live in our intestines and on other body sites were largely unknown and indeed were of almost no interest to the majority of people, including most scientists. The only real exceptions were organisms shown to cause diseases such as diarrhoea. The past 15 years, however, have seen an explosion of research into what has been called the 'human microbiome' and especially the micro-organisms that reside in our gut (the 'gut microbiota'). As soon as this research started to suggest links between previously unconsidered micro-organisms and human health, its importance was reflected in the media and began to be appreciated by the general public. Claims that our microbiome is linked to the prevention or causation of obesity, diabetes, cancer, bowel disease, allergy and autoimmunity, resistance to infection and even mental health have now become commonplace.

The explanation for this sudden explosion in 'microbiome research' is technological. Just as the invention of the microscope was fundamental in creating the field of microbiology, new techniques for the rapid sequencing of nucleic acids in the twenty-first century have revolutionised the study of microbial 'communities', bypassing the need to grow (culture) individual organisms in the laboratory. This advance quickly attracted a major wave of new research funding to what was now seen as a hot topic. Before this, it had been left to a small cadre of dedicated microbiologists and physiologists to study the gut microbiota and its effects on the host. As we will see, the major inhabitants of the gut microbiota are anaerobes that do not grow in the presence of air. This makes them inconvenient to study through culturing, and young microbiologists were easily put off working on anaerobes if they wanted to have a successful career (although on a personal note I am pleased to say that this did not deter me!). There are, however, some very important insights from this type of work that are in danger of being overlooked in the headlong rush towards sequence-based analyses.

This book aims to present a concise and balanced overview of what we currently understand about our gut microbiota. In doing so, it attempts to relate advances gained from the latest high-throughput sequence-based research to understanding that comes from other approaches, especially culture-based microbiology and systems modelling. This book sets out to explain each topic as far as possible from first principles. This is intended to make it accessible to non-specialist readers who have some basic background in science, but who might not have formal training in microbiology or biology. A Glossary is included at the end that may prove helpful. Selected references are included for those who wish to read further, although these represent a very small subset of the thousands of papers being published annually in this field.

Most of all, I hope that this book manages to communicate some of the excitement, fascination and promise of this rapidly developing research field. Some of this is conveyed in the artistic frontispiece by my daughter Rowan, inspired by the content of this book.

Aberdeen, UK   Harry J. Flint

Acknowledgements

I would like to thank my colleagues Petra Louis and Sylvia Duncan for taking the time to read and comment on these chapters while still in draft form. Their input was extremely valuable and certainly helped to improve the content. Thanks also to Petra for designing Fig. 4.1, to Alan Walker for allowing me to show the 'FISH' micrographs from his PhD thesis in Fig. 5.2 and to Helen Kettle for the modelling in Fig. 11.2. I am very grateful to another ex-colleague, Tim King, for his permission to show the excellent electron micrograph in Fig. 3.1. Very special thanks are due to my artist wife Irene who produced the diagrams of gut anatomy for Figs. 2.2 and 5.1 and to my artist daughter Rowan for the frontispiece. I would also like to take this opportunity to thank all of those who worked with me in the Microbiology/Gut Health group of the Rowett Institute in the University of Aberdeen over the years. Our research discussions and shared problem-solving undoubtedly helped to shape many aspects of this book, and I would like to acknowledge the Scottish Government (RESAS) for supporting our research. I must also pay tribute to my late PhD supervisor and mentor, Henrik Kacser, whose insistence on questioning and enquiry left an indelible impression on my approach to science.

I am grateful to Marcus Spaeth of Springer for his patient encouragement in getting this project off the ground. And to Alan Walker for the initial suggestion that prompted me to write this short book. I have very much enjoyed the project, and I hope that readers will find the topic as fascinating as I do. I have tended to avoid the more anecdotal and journalistic style that is to be found in some popular science books in this field. This is not out of any sense of scientific puritanism. It is more a feeling that the real beauty and meaning of science, as well as the critical importance of evidence and uncertainty when drawing conclusions, can easily get lost if the subject matter is over-simplified or trivialised. Here I have attempted to explain the science of gut microbiology in a way that I hope will be accessible to an interested layperson, whilst also offering a concise account to those with a more specialist interest in the field. What I have written is a personal account, and there may well be room for disagreement with some of the interpretations and views expressed. I have tried to avoid the pitfalls of overstating preliminary findings or overclaiming recent advances, but hopefully have at the same time highlighted some of the truly exciting progress that is being made in this field.

Huge thanks go to my wife Irene for her patience, support and encouragement over the past 28 years. Thanks also to my children, Kathryn, Christopher and Rowan for their encouragement and interest in my endeavours.

Contents

Chapter 1
Micro-organisms and the Microbiome

The term 'Micro-organism' refers to any living thing that is too small to be seen with the naked eye. It covers an astonishing array of life forms that began with the earliest living occupants of our planet. Of the three recognized domains of life, two (Bacteria and Archaea) consist exclusively of micro-organisms. The third (Eukaryota) contains all of the macroscopic, multi-celled organisms that we recognize as plants and animals, but it also includes many micro-organisms. 'Microbiome' is a collective term[1] for all of the micro-organisms belonging to these three domains, while the 'Biome' comprises all life on earth. For most of earth's history until the evolution of multicellular Eukaryota around 600 million years ago, however, the earth's Biome consisted only of micro-organisms (Fig. 1.1). Even now the Microbiome is estimated to represent more than half of the total living matter (*biomass*) on the planet [1]. Their invisibility makes it easy for us to overlook the vast impact that micro-organisms have on the sustainability of the planet and of life on earth. While this book will focus on the gut-associated Microbiome, it is important that we start with a look at the wider microbial world.

Bacteria and Archaea

Members of the two exclusively microbial domains of life, Bacteria and Archaea, appear superficially similar under the microscope, but they are recognised as distinct because of fundamental differences in the way that their cell structures and genetic material are organised [2]. Evolutionary divergence between the ancestors of the

---

[1]The term 'Microbiome' is used here to refer to the microorganisms themselves and is synonymous with '*microbiota*'. Somewhat confusingly, 'Microbiome' has also been used by some to refer to the collective genetic material of these microorganisms (synonymous with the term '*metagenome*', discussed later (Chap. 3)).

© Springer Nature Switzerland AG 2020
H. J. Flint, *Why Gut Microbes Matter*, Fascinating Life Sciences,
https://doi.org/10.1007/978-3-030-43246-1_1

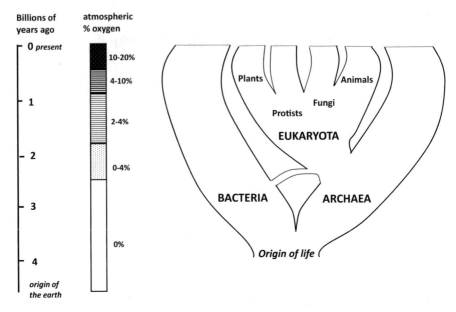

Fig. 1.1 Schematic of the three domains of life. Timescale and change in atmospheric oxygen are shown on the left

Archaea and Bacteria is thought to have occurred very early in the evolution of life on earth, at least 3.5 billion years ago.

All living organisms require a supply of the chemical elements of which they are made (notably carbon) together with a source of energy. This gives them the ability to make, or to obtain, the complex molecules that they require to grow and to reproduce themselves. Clearly the first cellular life forms on earth must have been able to gain their energy and nutrition without relying on any other living organism. Such organisms are called *autotrophs* and are still well represented within the present-day Microbiome. Some autotrophs are able to gain their energy from chemical reactions and their carbon from carbon dioxide or methane (by processes known as *chemosynthesis*). For example, bacteria found in hydrothermal vents deep in the ocean gain energy from the conversion of hydrogen sulfide to sulfur, and this process helps to support a complex food web that includes huge tube worms with which the micro-organisms are closely associated [3]. Indeed, recent evidence suggests that deep ocean vents are a possible site for the origin of life, with chemosynthesis as the driving force [4]. Other extremely important autotrophs, such as the Cyanobacteria, are able to gain energy from sunlight through the process of *photosynthesis*, while absorbing carbon dioxide and producing oxygen. Early evidence of life on earth is provided by 3.5 billion year-old fossilised mats of Cyanobacteria known as 'stromatolites' and it was the photosynthetic activity of these organisms over a period of two billion or more years that first introduced oxygen into the earth's atmosphere. Even now, the marine cyanobacterium *Prochlorococcus* and its close relatives are estimated to account for 50% of global consumption of atmospheric carbon dioxide by the oceans [5]. By no means all photosynthetic bacteria produce oxygen. Autotrophic green sulfur and purple sulfur bacteria (*Chlorobium, Chromatium*) are photosynthetic organisms that are able to use hydrogen or reduced sulfur compounds to make their cell components from carbon dioxide, meaning that photosynthesis can occur in environments (both now and in the distant past) that lack oxygen. As shown in Fig. 1.1, atmospheric oxygen levels were minimal or very low during the first 3 billion years of microbial evolution.

The Archaea include some truly remarkable organisms, many of them autotrophs. Among them are the only living things that are able to gain energy by forming methane gas (*methanogens*). An extraordinary square-celled archaeon (*Haloquadratum*) that is able to use sunlight for energy exists in salt pans in salt concentrations 10 times that of sea water. Other archaeal '*extremophiles*' (organisms that prefer extreme environments) are able to exist at the temperature of boiling water in hot springs (*Pyrococcus furiosus*), or at high pressures found in the deep ocean, or at extremes of acidity and alkalinity. Extremophiles are also found among the Bacteria, for example *Deinococcus radiodurans* which has the ability to withstand extremely high levels of radiation. It is a reasonable assumption that extremophiles evolved initially under the extreme environments that prevailed during the early life of the planet, but they continue to survive today in extreme habitats such as saline lakes and hot springs. Other groups of Archaea play very important roles in the environment by oxidising methane and ammonia.

The great majority of present-day micro-organisms are not autotrophs that feed themselves, but *heterotrophs* that gain their energy and carbon by making use of compounds derived from other living organisms (i.e. *organic* compounds). Leaf litter, soils and sediments contain huge quantities of organic matter that provide abundant nutrients for microbial growth. The highest concentrations of micro-organisms in nature are found in the intestinal tracts of animals, particularly herbivores, where they benefit from the availability of plant material that is not digestible by the host animal [1]. As we will see later, such associations are often mutually beneficial (*symbiotic*), with the host able to benefit from the products of microbial activity and the micro-organisms benefitting from a stable environment and guaranteed food source. Important symbiotic interactions also occur between plants and microorganisms, for example in the nitrogen-fixing root nodules of clover. On the other hand, microbial interactions with multi-cellular plants and animals can also be very one-sided, and we are all too familiar with the fact that particular bacteria (termed *pathogens*) can cause serious diseases in their hosts. Archaea have not so far been implicated directly in disease causation.

While most bacteria and archaea exist as single cells, others grow as long branching strings (filaments). In some cases, more complex organisation occurs within filaments through branching or cell specialisation. Many bacterial species have been shown to form spores that enable them to survive high temperatures and adverse chemical environments (e.g. alcohol) that kill the normal (vegetative) cells. Such resistant spores can lay dormant for very long time periods. Spore-formers that are pathogens (e.g. anthrax, several *Clostridium* species) are of particular concern as we require extreme chemical treatments or the high temperatures and pressures created through *autoclaving* to destroy them and so ensure sterility. A salutary example is provided by Gruinard Island in northwest Scotland, which was used for the experimental release anthrax spores in 1942. 48 years later the island was finally declared safe, but only after treatment of the entire 196 hectare surface with formaldehyde and removal of much of the topsoil. Unlike many Bacteria, Archaea are not known to form spores.

Eukaryota

Diverse single-celled organisms related to algae, fungi and protists are also referred to as micro-organisms, but because their cells show a complex internal organisation that is absent from bacterial and archaeal cells they clearly belong to the third domain of 'Eukaryota' (often called 'eukaryotes'). In particular, their genetic material is contained within a membrane-enclosed *nucleus* and they possess other specialist *organelles* such as *mitochondria* within the cell. The first eukaryotic cells are believed to have evolved around 2 billion years ago. One of the most astonishing conclusions of evolutionary biology is that these first eukaryotic cells apparently originated from an intimate co-operation between an archaeal precursor cell and a bacterial cell. Thus, the energy-generating organelle (the mitochondrion) that is found within the cells of almost all eukaryotes was originally derived from a free-living bacterial cell, while other cell components apparently owe more to the distant archaeal predecessor. Mitochondria play a vital role in the ability of eukaryotic cells to gain energy via respiration. It is also considered that the *chloroplast* structure that is responsible for photosynthesis within the eukaryotic cells of algae and green plants was originally a symbiotic cyanobacterium. Thus the basic machinery required for photosynthesis and respiration in plants and animals had evolved within the Microbiome long before the evolution of macroscopic life forms.

Our view of the relatedness of different single-celled eukaryotes has changed drastically over the past 50 years. Most fungi are now considered as close to animals as to plants in their evolutionary origins. Single-celled organisms once referred to collectively as 'protozoa' (such as flagellates, amoebae and ciliates) are now considered to belong to multiple, well separated branches on the evolutionary tree, and the same is true for different groups of photosynthetic algae (diatoms, red, green and brown algae). It follows that multicellular life forms must have evolved not once, but many times over in different evolutionary lineages. There is an incredibly wide range of growth habits among eukaryotes intermediate between the single-celled state and a truly multicellular state that involve varying degrees of organisation, aggregation, communication and specialisation, as illustrated by slime moulds, algal chains and colonies. Many fungi grow as microscopic filamentous *hyphae* that form an extensive network, or *mycelium*. On the other hand many of them also develop complex structures that make it hard to describe them as micro-organisms, suddenly appearing as large, highly visible fruiting bodies (mushrooms). Likewise, the red, green and brown algae that we call seaweeds can clearly exist as very large and complex structures. Nevertheless, many single-celled eukaryotes (such as yeasts and amoebae) are truly microscopic and are quite properly counted as members of the Microbiome. Eukaryotic micro-organisms include many important plant and animal pathogens, both among the fungi and protists. Most of them are non-pathogens, however, that play extremely important roles in the environment through photosynthesis and in the recycling of organic material.

The Virome

Viruses are entities that reproduce and propagate themselves by subverting the reproductive machinery provided by archaeal, bacterial or eukaryotic cells. Although they must therefore have evolved after the first living cells, the nucleic acid relationships between viruses and other life forms indicate an early origin. In fact, viruses are strictly non-living structures since they consist only of genetic material (which can be either DNA or RNA depending on the virus) in a protective coat. They are referred to collectively as 'the Virome'. Their impact on the living organisms of the Biome is, however, profound. We are most familiar with viruses as agents of human diseases ranging from chickenpox and influenza to ebola and coronavirus. Viruses are also responsible for many diseases in crop plants, and indeed tobacco mosaic virus (TMV) was the first virus of any kind to be studied in detail. But there are also huge numbers of viruses that infect bacteria and archaea, and these play a major, if still little understood, role in influencing the cell populations and evolution of their target micro-organisms.

Symbiosis

The huge significance of symbiosis between living organisms is a recurring theme in biology. We will use the term symbiosis here to mean 'living together for mutual benefit' (also known as 'mutualism') as distinct from a situation in which one partner benefits and the other comes to no harm from the relationship (*commensalism*). Many important and fascinating symbioses occur between multi-cellular plants or animals and microorganisms. A particularly close interaction occurs between leguminous plants such as clover and specific bacteria, with the plant's roots forming special nodules that house the bacteria (often *Rhizobium* species) responsible for converting ('fixing') atmospheric nitrogen to form ammonia [6]. As a result, the plant gains a supply of usable nitrogen that does not rely on nitrogenous compounds being supplied from the soil. This symbiotic relationship has been, and continues to be, of enormous significance for agricultural production as it reduces our reliance on chemical fertilisers.

Another vital symbiotic relationship occurs between land plants and fungi. It is estimated that 70–90% of land plant species are involved in associations between their root systems and fungi called *mycorrhizae* [7]. Indeed, just a cubic millimetre of soil can contain hundreds of metres of mycorrhizal fungal hyphae. These extensive fungal networks improve uptake of minerals and of phosphate by the plant, while giving the microbial partner a protected environment and access to photosynthetically produced energy in the form of sugars.

We will look at symbioses between animal hosts and gut micro-organisms in the next chapter. One particularly intriguing association is worth mentioning here, however, that illustrates a symbiotic association between an animal and a non-gut micro-organism. Many marine organisms are bioluminescent—that is to say, they emit light. In the Hawaiian bobtail squid, the ability to emit light is entirely due to a bacterial symbiont, *Vibrio fischeri*, which possesses the light producing-enzyme luciferase. The squid manages to select this symbiont from the environment very early in its development and allows it to proliferate within a special light organ, thus giving it the ability to glow in the dark [8]. *Bioluminescence* is thought to have a variety of benefits for ocean dwellers, but in this case, it is suggested that the squid is camouflaging itself from predators in shallow water by replacing the natural light blocked by its body with bioluminescence!

Many other symbioses are known to occur between microorganisms themselves. A spectacular example is provided by lichens, which are associations between two or more fungal partners and one or more green algae or cyanobacteria. As the photosynthetic partners, the algae or cyanobacteria produce carbohydrates by using light energy, and these in turn become available as energy sources to the fungal partner. Meanwhile the fungal network of filaments (hyphae) helps to anchor the photosynthetic cells and protect them from desiccation. This association is extremely successful, since lichens are well known as the first pioneering colonisers of barren surfaces from rocks to tree bark. Although composed of micro-organisms, lichens are visible as macroscopic structures that are often highly coloured and structurally quite complex. The relationship is clearly symbiotic, although the precise balance of benefit between the partners is likely to vary widely among the 20,000 lichen species that have been described!

Nutrient Cycling

Almost any organic (carbon-based) molecule that is made by one organism can be made use of by another organism within the Microbiome under some environmental conditions. As a result, there is very efficient carbon recycling and this tends to limit the accumulation of carbon-based molecules on the earth. Massive accumulations do occur, however, both in the present and over geological time—peat bogs represent the accumulation of organic matter when soil conditions are too waterlogged and acidic to allow the normal breakdown by micro-organisms. Most obviously, fossil fuels such as oil and coal are the product of un-degraded carbon that was deposited and then subjected to enormous temperatures and pressures resulting from geological upheaval hundreds of millions of years ago.

Microbial activity is not only crucial to the carbon economy of the planet, but also to that of most of the major elements found in living things. Atmospheric nitrogen is converted into soluble forms of nitrogen (nitrate, ammonia) by nitrogen-fixing bacteria which can be found in root nodules of many plants, and also by other nitrogen-fixing micro-organisms that are free-living. Since animals and plants cannot use gaseous nitrogen, this activity provides the usable nitrogen needed to make their proteins and nucleic acids. Conversely, ammonia oxidation by bacteria and archaea leads the formation of gaseous nitrogen, leading to a complex cycle of nitrification and de-nitrifcation.

How Many Microbiomes?

The remainder of this book will focus increasingly on the microbiota of the animal and human gut, but we have started with consideration of the global Microbiome for good reasons. Micro-organisms are incredibly numerous and occur in almost all environments found on the surface of the planet. They are capable of dissemination pretty much anywhere via the atmosphere, water and wind and via biological vectors, excreta and fluids. As a result they do not respect boundaries and the term 'Microbiome' is best applied to the whole planet. The soil that we walk on, the air that surrounds us and that we breathe, the food that we eat and the surfaces that we touch are teaming with all kinds of micro-organisms and their spores. While reference is often made (including in this book) to the 'Gut Microbiome' or the 'Skin Microbiome', we should always remember that this is just part of the global Microbiome. Certain human gut micro-organisms, including some pathogens, are capable of surviving and propagating themselves in other host species, or indeed outside the gut. Others appear more host-specific, but still have to be transmitted between individuals, especially between parents and offspring. More fundamentally, we should not forget that the evolutionary origins of gut micro-organisms lie with the ancient microbial world that existed long before the emergence of multicellular life forms and the development of digestive tracts.

References

1. Whitman WB et al (1998) Prokaryotes: the unseen majority. PNAS 95:6578–6583
2. Woese CR, Fox CE (1977) Phylogenetic structure of prokaryotic domain – primary kingdoms. PNAS 74:5088–5090
3. Cavanaugh CM et al (1981) Prokaryotic cells in the hydrothermal vent worm *Riftia pachyptila* Jones – a possible chemoautotrophic symbiont. Science 213:340–342
4. Martin W et al (2008) Hydrothermal vents and the origin of life. Nat Rev Microbiol 6:805–814

5. Fu F-X et al (2007) Effects of increased temperature and CO_2 on photosynthesis, growth and elemental ratios in *Synechococcus* and *Prochlorococcus* (Cyanobacteria). J Phycol 43:485–498
6. Sugawara M, Sadowsky MJ (2011) Legume-microbe symbioses. In: Rosenberg E, Gophna U (eds) Beneficial microorganisms in multicellular life forms. Springer, Berlin, pp 73–88
7. Minz D, Ofek M (2011) Rhizosphere microorganisms. In: Rosenberg E, Gophna U (eds) Beneficial microorganisms in multicellular life forms. Springer, Berlin, pp 105–121
8. Nyholm SV, McFall-Ngai M (2004) The winnowing: establishing the squid-vibrio symbiosis. Nat Rev Microbiol 2:632–642

Chapter 2
The Gut Microbiome: Essential Symbionts or Unwelcome Guests?

Nearly all animals possess a digestive tract (the exception being parasites that live within their host's tissues or digestive tracts) and in nearly all cases this consists of an open tube with flow of contents from one end (mouth or stoma) to the other (anus). Since the animal is ingesting food from its environment, it is also ingesting all of the microbes that are associated with the food, along with those from the immediate environment. This means that, although the gut within a foetus developing in the uterus is normally sterile, the adult gut has no chance of being sterile. It will inevitably become colonised by micro-organisms that are able to take advantage of the readily available sources of energy to be found in the gut.

In view of this, what is remarkable is that each animal species is found to have a characteristic, although usually highly diverse, collection of micro-organisms within its gut [1]. Indeed, it is very common to find that particular microbial species are associated almost exclusively with particular animal host species. This suggests that extensive evolution of the gut microbiota (often resulting in the formation of new species) has occurred through adaptation to the particular host. In fact, certain gut micro-organisms appear to 'belong' to their host (the term '*autochthonous*' has been used for such resident microbes) [2]. We can see therefore that the gut microbiota of a given host is not simply a random collection of organisms that happen to have come into the gut with the food supply, but a special set of organisms that have become co-adapted to each other and to their host. Other organisms enter the gut continually throughout the animal's lifespan, but most of these will pass through without reproducing themselves or reaching significant numbers (these are called '*allochthonous*' or, put more simply, 'transient' species) [3]. Of the many types of interaction that occur between the host and micro-organisms within the gut, it is symbiotic relationships that underpin the largest and nutritionally most significant microbial communities. We will have a quick look at two of the best studied examples of animal host-gut microbe symbiosis.

© Springer Nature Switzerland AG 2020
H. J. Flint, *Why Gut Microbes Matter*, Fascinating Life Sciences,
https://doi.org/10.1007/978-3-030-43246-1_2

Termites

It should come as no surprise to find that the digestive tracts of insects are colonized by micro-organisms. Interactions between the insect and its microbiota however are often intriguing and complex. Many termites (relatives of cockroaches) have a remarkable ability to digest wood, which is why they are a serious threat to timber buildings and furniture in tropical countries. Unusually among animals, these insects produce their own enzymes that contribute to the breakdown of cellulose, which along with lignin is the major constituent of wood. It appears however that their own termite cellulases are not sufficiently active to allow them to survive on a diet of wood. In 'lower termites' the breakdown of wood fragments requires the activity of symbiotic micro-organisms, specifically flagellated protists (often just called 'flagellates') located within their hindgut, and without these the insects starve [4]. The flagellates contribute additional cellulase enzymes that help to break down the woody fragments (*lignocellulose*) within their food *vacuoles* (fluid-filled spaces within the cell that could be considered analogous to a stomach). The *fermentation* of sugars derived from the breakdown of cellulose, which occurs under conditions of low oxygen in the insect's gut, mainly produces acetic acid (acetate)[1] (Fig. 2.1). The insect then absorbs and uses the acetate as a source of energy, so indirectly gaining access to calories derived from the cellulose in wood. In addition, the flagellates are themselves associated with bacteria and with methanogenic archaea, some of which are on the cell surface and some (as *endosymbionts*) are actually located within the cells of the flagellate. Another problem with relying on wood as the main food is that it has a very low nitrogen content. While the gut microbes help to recycle nitrogen from the insect's proteins and nucleic acids via uric acid, a more crucial role is played by bacterial symbionts that fix and reduce atmospheric nitrogen, thus helping to compensate for the low dietary intake of nitrogen [6]. Meanwhile the most abundant group of bacteria in the termite gut, the Spirochaetes, convert much of the hydrogen that is released during cellulose breakdown into acetate (by the process of reductive acetogenesis (see Chap. 4)), thus boosting the supply of acetate to the host. An alternative fate for hydrogen is its conversion to methane gas by the methanogenic archaea, but the more central location of the *acetogenic bacteria* within the termite gut is thought to favour acetate formation in many species.

'Higher termites' (which include the ones most familiar to us, for producing termite mounds) have apparently lost their flagellates at some point during their evolution. Higher termites do however contain large numbers of bacterial symbionts in their gut, including not only Spirochaetes but also Fibrobacteres, a group that is important in cellulose breakdown in the rumen. Presumably it is these symbiotic bacteria, rather

---

[1]Acids dissociate readily to release a hydrogen ion (H^+). When acetic acid releases H^+, it becomes the 'acetate' anion (A^-). Both the dissociated and non-dissociated forms exist together, with their proportions depending on the prevailing pH. For convenience we will generally refer to the anion (formate, acetate, lactate etc.) in this book.

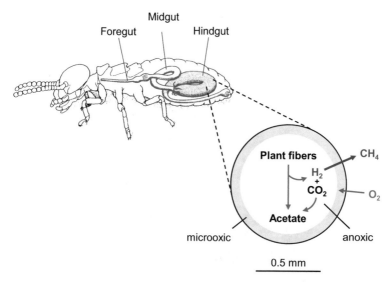

Fig. 2.1 Termites depend on symbiosis with gut micro-organisms to use ligno-cellulose (wood). The termite gains energy by absorbing the acetate produced by fermentation of cellulose, while some carbon is lost in combination with hydrogen (H_2) as methane (CH_4). Some oxygen (O_2) diffuses in from the blood, but the contents of the hindgut have little or no oxygen ('microoxic' or 'anoxic'). From—Andreas Brune [5]

than flagellated protists, that complement the activity of the termite cellulases to achieve the breakdown of lignocellulose from wood in these higher termites.

Herbivorous Mammals

The termite gut microbiota has many parallels with the anaerobic breakdown of lignocellulose by symbiotic micro-organisms in the ruminant gut. Indeed, the termite gut was first chosen for study, along with the rumen, by Robert Hungate in his pioneering research on microbial symbiosis in the animal gut [7]. Unlike termites, mammals do not produce their own cellulases and so are wholly reliant on micro-organisms within their gut for any ability to breakdown the cellulose and plant fibres that are in their diet. As a result, we find that in all herbivorous mammals that specialise in eating diets made up predominantly of grass and other fibrous plant material, the gut has evolved to encourage fermentation by microbial symbionts. There are two basic models, 'pre-gastric' and 'hindgut' (i.e. post-gastric) fermentations [8].

In pre-gastric fermenters, which include ruminants such as cattle, sheep, goats, deer, antelope, camels and llamas, food goes straight from the mouth via the oesophagus into a set of large fermentation chambers. There it is subjected to

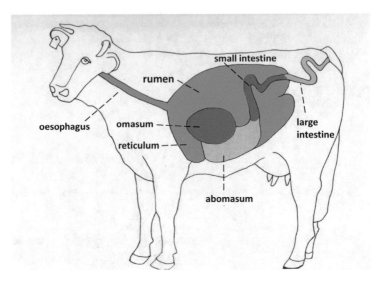

Fig. 2.2 The ruminant gut

extensive fermentation by the resident microbial community at a neutral pH (neither acidic nor basic, close to pH 7).[2] Cows are traditionally said to have four 'stomachs', but this is slightly misleading. Rather, they possess a single highly complex organ that is made up of four interconnected chambers. The first two, known to anatomists as the 'reticulo-rumen' (but generally shortened to rumen) are the main sites of fermentation and have the largest volumes (Fig. 2.2). From these, the digesta (i.e. food that is in the process of digestion) enter a smaller chamber called the 'omasum' where there is rapid absorption into the bloodstream of water and of the acidic products of microbial fermentation. Absorption also occurs across the walls of the rumen itself, and indeed camels and llamas (sometimes called 'pseudo-ruminants') manage with only three chambers, as they lack an omasum altogether.

  After this, the digesta enter an acidic chamber called the 'abomasum', which is roughly equivalent to our stomach. What happens here is a unique feature of the pre-gastric model. Huge numbers of microbial cells flow into the abomasum along with the digesta leaving the omasum (or leaving the rumen, if you are a camel). These microbial cells, which were of course the ones responsible for fermenting the dietary plant material in the rumen, are then themselves digested in the acidic abomasum. This requires the ruminant to produce a specially adapted *lysozyme*, an enzyme that is responsible for cracking open the cell walls of bacteria. Whereas other mammalian lysozymes, such as human salivary lysozyme, operate at a fairly neutral

---

[2]The pH is a number that represents the acidity or alkalinity of the environment on a scale of 0 (most acidic) to 14 (most alkaline (or basic)). pH 7 is defined as neutral. pH is determined by the concentration of hydrogen ions; specifically, pH is the negative logarithm of the hydrogen ion concentration (e.g. a H^+ concentration of 10^{-5} M is pH 5).

Table 2.1 Fermentative capacity of the gut in different mammals (based on van Soest [8], Parra [10])

Fermentative capacity (as % of total digestive tract)

Animal	Reticulorumen	Cecum	Colon and rectum	Total
Sheep	71	8	4	83
Cattle	64	5	5–8	75
Capybara	–	71	9	80
Horse	–	15	54	69
Rat	–	32	29	61
Rabbit	–	43	8	51
Pig	–	15	33	48
Human	–	–	17	17
Cat	–	–	16	16
Dog	–	1	13	14

pH, the lysozymes secreted into the bovine abomasum has evolved to operate at an acidic pH and also to withstand the digestive enzyme pepsin, whose role is to break down proteins in the stomach [9]. The huge benefit this has for the ruminant animal is to provide a much-needed supply of nitrogen. While the plant material consumed in the diet contains some protein, the rumen microbiota includes a great many species that break down this dietary protein and use the nitrogen supply for their own growth and metabolism. In effect, they are stealing the supply of dietary nitrogen from the animal host. So, the ruminant simply gets this nitrogen back by consuming its own microorganisms in the abomasum! One might argue that it is a strange form of symbiosis, in which one partner gets destroyed after making its contribution, but of course the rumen microbiota continue to flourish and to reproduce in a uniquely stable, nutrient-rich environment within the host.

The sheer volume of these fermentation chambers is extraordinary. In an adult cow, the reticulo-rumen and omasum can be 50 litres in volume and account for one sixth of the total volume of the body, or three-quarters of the abdominal cavity [10] (Table 2.1). This represents a huge investment by the animal in the harvesting of energy and nutrients from microbial activity. Several other intriguing cases of pre-gastric fermentation are known among animals. These include a group of leaf-eating colobine monkeys, which like ruminants produce an acid-resistant lysozyme that allows them to digest their gut microorganisms. Other examples of non-ruminant pre-gastric fermenters are kangaroos and the hippopotamus. More unexpected is the hoatzin, a South American bird, also known as the 'stink bird', where microbial fermentation of plant matter occurs in the crop, releasing the smelly volatile acids that give the bird its popular name.

Hindgut fermenters represent a more diverse group of animals with a wider range of nutritional habits. They do however include many large herbivores that (like cattle and sheep) spend much of their time grazing, including horses, zebra, elephants and capybara. Here, microbial fermentation occurs in an enlarged caecum and/or colon (sections of the large intestine) after the digesta have passed through the acidic stomach and the small intestine. It is estimated that almost 70% of the total volume of the horse intestine is devoted to microbial fermentation of plant material, a figure only slightly less than the 75% reported for cattle [10].

The range of eating habits found among different species of mammal is reflected in, and is largely determined by, their digestive anatomy. Herbivores such as rabbits that feed selectively on more digestible types of plant material, show a substantial capacity for hindgut fermentation somewhat less than that of grazing herbivores. Nevertheless, even omnivores such as pigs, mice and rats still devote around 50–60% of their gut volume to caecal and colonic fermentation. Many hind-gut fermenters (including rabbits and rodents) maximise the energy that they extract from the diet by eating their own faeces, thus ensuring that there is a second passage of the digesta through the hind-gut (this habit is known as *coprophagy*). Only when we come to carnivores such as dogs as cats do we find that the large intestine accounts for a relatively small fraction (less than 20%) of the gut volume. Where do humans sit in this range? The answer (17%) might surprise some people, in that we sit firmly with the carnivores. Nevertheless, this may be consistent with the hunter-gatherer lifestyle that we assume was followed through the larger part of human history and prehistory.

We will look at the activities of the rumen microbiome in more depth in Chap. 4 and at our own human microbiome in Chap. 5 onwards.

References

1. Ley RE et al (2008) Evolution of mammals and their gut microbes. Science 320:1647–1651
2. Nava GM, Stappenbeck TS (2011) Diversity of the autochthonous colonic microbiota. Gut Microbes 2:99–104
3. Maldonado-Gomez MX et al (2016) Stable engraftment of *Bifidobacterium longum* AH1206 in the human gut depends on individualised features of the resident microbiome. Cell Host Microbe 20:515–526
4. Brune A, Carsten D (2016) The gut microbiota of termites: digesting the diversity in the light of ecology and evolution. Annu Rev Microbiol 69:145–166
5. Brune A (2011) Microbial symbioses in the digestive tract of lower termites. In: Rosenberg E, Gophna U (eds) Beneficial microorganisms in multicellular life forms. Springer, Berlin, pp 3–25
6. Hongoh Y et al (2008) Genome of an endosymbiont coupling N_2 fixation to cellulolysis within protist cells in the termite gut. Science 322:1108–1109
7. Hungate RE (1966) The rumen and its microbes. Academic, New York
8. Van Soest PJ (1994) Nutritional ecology of the ruminant, 2nd edn. Comstock Publishing Associates, Cornell University Press, Ithaca and London
9. Irwin DM (1995) Evolution of the bovine lysozyme gene family: changes in gene expression and reversion of function. J Mol Evol 41:299
10. Parra R (1978) Comparison of foregut and hindgut fermentation in herbivores. In: Montgomery GG (ed) The ecology of arboreal folivores. Smithsonian Institution Press, Washington DC, p 205

Chapter 3
How to Analyse Microbial Communities?

Most micro-organisms exist in nature as complex mixtures which are referred to as microbial communities. Members of such communities potentially interact with each other in various ways, including as competitors, predators and symbionts. Communities range from extremely dense mixtures of organisms that are held together in biofilms by binding to each other and to a surface (e.g. in dental plaque), to the highly dilute communities that are found in nutrient-poor ocean water. Finding out what organisms are present within a microbial community and how they interact with each other is one of the most intriguing challenges in biology.

Microscopy

The most direct way to observe the micro-organisms that comprise a community is to look at them down a microscope. Viewing a live sample of rumen fluid under a light microscope at low magnification, for example, reveals the extraordinary sight of numerous large cells with complex surface features that are in constant motion across the field of vision. These are ciliated rumen protists, which range from 20 to 500 microns (0.02–0.5 mm)[1] in length. Higher magnification is needed to reveal bacterial cells, which are more typically in the range 0.5–5 microns. Classical Gram staining can distinguish between cells with different types of bacterial cell wall, with Gram-positive cells staining purple and Gram-negative cells pink (although retention of the stain is also affected by growth stage). It is possible to perform an approximate total cell count by microscopy, and stains can be used that differentiate live from dead cells. In the end, however, one is confronted with a bewildering mixture of cells of differing sizes and shapes that provide little clue as to their function or role in the community!

---

[1] A micron is one thousandth of a millimetre.

© Springer Nature Switzerland AG 2020
H. J. Flint, *Why Gut Microbes Matter*, Fascinating Life Sciences,
https://doi.org/10.1007/978-3-030-43246-1_3

Fig. 3.1 Rumen microorganisms attacking a leaf fragment. Transmission electron micrograph of a section through a sample of rumen contents (courtesy of Dr. Tim King)

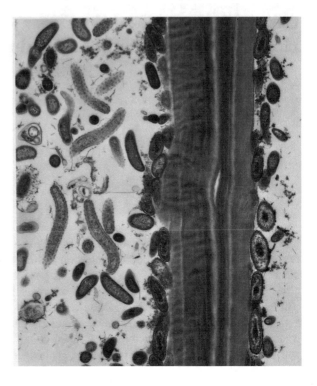

Fortunately, more sophisticated approaches are available using fluorescent '*probes*', based on short nucleic acid sequences or on antibodies, in conjunction with fluorescence microscopy, to highlight particular species or groups of microorganisms. Probes have also been used to physically sort, and then count, cells in an automated manner by the technique known as fluorescence activated cell-sorting (FACS). In general, however, microscopy and cell sorting are simply too time-consuming to be the methods of choice for routine analyses of microbial community composition. As we will see shortly, much faster and increasingly powerful molecular methods are available for this purpose. Nevertheless, an enduring strength of microscopy is that it offers unique insights into the organisation of microbial cells and their relationships to each-other, to food particles and to host cells (Fig. 3.1). This requires sophisticated techniques such as histochemistry, confocal microscopy and electron microscopy. The very highest magnifications are achievable through transmission electron microscopy, which allows visualisation of details of cell surfaces and internal structures down to single DNA molecules and enzymes.

Culturing Micro-organisms

The question of what the many organisms that comprise a microbial community actually live on, and what their activities are, has traditionally required their isolation and cultivation in the laboratory. An artificial 'medium' that encourages the growth of many micro-organisms can be prepared quite easily from nutrient-rich ingredients such as yeast extract, hydrolysed milk protein (casein) and carbohydrates. The medium can be solidified by addition of agar (an inert gel-forming material made from seaweed). This creates a solid surface after pouring the molten liquid mixture into sterile plastic (originally glass) lidded dishes (petri dishes) and allowing it to cool. Media are sterilised during preparation, typically by autoclaving (also used for sterilisation of surgical material in hospitals), so that only microbes added by the experimenter will grow.

If a small amount of non-sterile liquid is spread over the agar 'plate' surface, this will give rise to noticeable microbial growth when left overnight at a moderate temperature. Often this growth will appear as a fuzzy continuous layer if the numbers of microbes in the sample are at all large (i.e. several hundred or more). If the sample is diluted sufficiently in a sterile solution and then plated, isolated colonies will form, with each colony being the product of a single microbial cell. Individual colonies can then be carefully picked off the plate and used to start 'pure cultures' that have been derived from a single strain (although in practice it is advisable to re-isolate each one several times to ensure purity). This is the principle behind most of the isolation methods that have been used to recover single micro-organisms from environmental samples for the past 100 years and more.

Samples of gut contents from the rumen or human large intestine contain extremely large numbers of micro-organisms—as many as 10^{11} (one hundred billion) cells per millilitre. This is clear from the numbers that can be seen and counted under the microscope. The plating procedure described above, if done in air, however typically results in numbers of viable colonies some 100–1000 times lower than this. In contrast, if the plating is done in the absence of air (in an atmosphere of carbon dioxide and, or, nitrogen) then the viable count can be well over 10^{10} per ml, which tells us that more than 99% of the micro-organisms in these samples are unable to grow in the presence of oxygen. Techniques that allow the isolation and culture of such *obligate anaerobes* were first developed in the late 1940s and 1950s, notably by Robert Hungate and Marvin Bryant in the USA, and in Europe by Kaars Sijpestein in the Netherlands and by others in France and the UK. In order to grow really 'strict' anaerobes, media have to be prepared in such a way as to exclude oxygen and then inoculated and incubated anaerobically. With less strict anaerobes that have some tolerance of oxygen it is possible to get away with plating in air, followed by incubation in an anaerobic atmosphere. For strict anaerobes, anaerobic glove boxes are available that allow plating to be performed in the absence of oxygen, generally in a mixture of carbon dioxide, nitrogen and hydrogen. An alternative to plating, the 'roll tube', was developed early on, in which molten agar and live cells are quickly solidified over the inner surfaces of a glass tube containing an atmosphere of carbon dioxide, by rotating the tube in cold water. These methods are all relatively time-consuming however and rather few laboratories routinely work with strict anaerobes. As a result, we rely on a handful of laboratories for our knowledge of such micro-organisms in culture.

Considerable effort went into isolating the major microbes that could be cultured from the rumen and human large intestinal microbiota over several decades from the 1950s onwards, resulting in the description of many new species of anaerobic bacteria for the first time. At the time, classifications were based mainly on the appearance of cells under the microscope (morphology), on the media and energy sources that would support growth and on the main products (acids, alcohols and gases) made by isolated cultures during growth. As we will see shortly, gene sequences play an increasingly important role in classifying micro-organisms, for the very good reason that they are more likely to provide an accurate reflection of the evolutionary (or 'phylogenetic') relationships between micro-organisms. Put simply, there is no particularly good reason to suppose that all bacteria with round ('coccus'-shaped) cells or the ability to produce lactic acid should necessarily be closely related to each other. On the other hand, we know that DNA sequences change progressively with the passage of time. These changes are driven both by natural selection and also by the process of genetic 'drift' that applies to mutational changes that may not affect evolutionary *fitness*. We can take it that two organisms with very similar DNA sequences across their genomes are generally more closely related to each other (i.e. they have diverged more recently from a common ancestor) than are two organisms with widely different genomic DNA sequences. The sequences of single genes common to all organisms (such as those coding for ribosomal RNAs), or of whole genome sequences, can be used to infer phylogenetic relationships between micro-organisms that represent their divergence through evolution.

Descriptions of previously cultured species cannot always be connected with present day sequence-based classifications. For one thing, many of the original 'type' strains have been lost. New anaerobic isolation work and approaches to isolation therefore have much to contribute by allowing sequencing of isolated strains to be combined with description of their growth and biochemistry. Such work is playing an important role in describing the diversity of human gut communities [1]. A long-standing approach for recovering micro-organisms that perform a specific function is selective isolation. For example, if we want to recover organisms that use a particular carbohydrate for growth, then we simply use a medium in which that carbohydrate is the only energy source added. This is how the small number of known cellulose-degrading bacterial species were first isolated from rumen contents and human faeces, thus allowing their identification [2, 3]. This approach remains highly relevant and has led, for example, to recovery of the important mucin-degrading bacterium *Akkermansia muciniphila* from human stool [4].

We should note that for some purposes, especially the detection of many pathogens, selective plating on media that are designed to recover only the organism of interest remains the most sensitive and definitive method of detection and is crucial for clinical laboratories.

DNA Sequencing and Culture-independent Analysis

Non-culture-based (or *culture-independent*) analysis has had a huge, indeed revolutionary, impact on microbial ecology over the past 25 years. Whereas 'plate counts' based on culturing were at one time the mainstay for microbial ecologists intent on describing microbial communities, they have now been supplanted almost entirely by molecular (DNA sequence-based) methodologies.

Living cells contain many essential proteins whose role is to process their genetic material *(DNA)*. Faithful DNA replication, which has to happen before cells can divide, is achieved by the enzyme *DNA polymerase*. DNA consists of two strands of *nucleotides* that are zipped together into a double helix by relatively weak forces ('hydrogen bonding'), but in a highly specific manner. Thus, of the four possible nucleotide bases, T (thymine) can pair only with A (adenine), and G (guanine) only with C (cytosine). Following separation ('melting') of the two strands of the DNA helix, DNA polymerase 'knits' the four nucleotides together into chains, with their sequence in the new strands being wholly determined by the pre-existing strands of the DNA helix that are used as templates. As a result, the two 'daughter' DNA helices each consist of one new strand and one old strand that precisely reproduce the sequence of nucleotides in the original (parent) helix (Fig. 3.2).

Elegant laboratory work showed that this complex cellular process can be reproduced in the test tube. Thus, given purified DNA and four *nucleoside* tri-phosphates (dNTPs),[2] new helices can be made using purified DNA polymerases. A brilliant and highly successful (although now largely historical!) approach to determining the nucleotide sequence of DNA molecules was also invented that depended on including artificial variants of each nucleotide that blocked further polymerisation by DNA polymerase (called 'dideoxy chain termination', or 'Sanger sequencing', after its inventor) in four separate reaction mixes. Following separation of the products of these aborted reactions by size (using radioactivity or fluorescence to detect them) the sequence of the original DNA molecule could simply be 'read-off'. Since the DNA polymerase requires a double-stranded section of DNA to start making the new strand, a short single-stranded primer sequence can be used to precisely define its start point. This technique made it possible to sequence individual genes, and then whole genomes, with the bacterium *Escherichia coli* being among the first to be completed [5]. The requirement for pure DNA as the starting material was generally met by *cloning* segments of DNA (using *recombinant DNA techniques*) for use as templates. The availability of genome sequence information from cultured micro-organisms has made a huge contribution to our understanding of their biology. But how can we relate this information to complex microbial communities *in vivo*? Also, what about species and strains that may not

---

[2]'dNTPs' refers to nucleoside triphosphates carrying deoxyribose (dATP, dTTP, dGTP, dCTP) which are the precursors for DNA synthesis. Two phosphate groups are lost as each nucleotide is added to the chain.

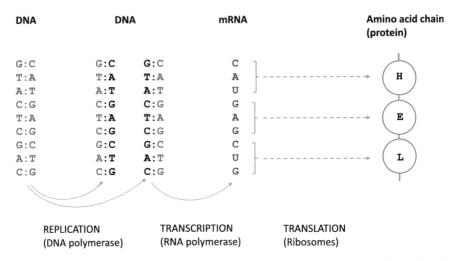

Fig. 3.2 How sequence information flows from the genome (DNA) to protein. G guanine, C cytosine, A adenine, T thymine, U uracil, H histidine, E glutamic acid, L leucine

have been cultivated? As we will see shortly, further developments in molecular techniques are allowing these awkward questions to be answered.

Polymerase Chain Reaction (PCR) Another revolutionary method involving DNA polymerase, the *polymerase chain reaction* (PCR), was invented in the early 1990s that allows the highly specific amplification of defined stretches of DNA. For this, primers that recognize matching sequences on the complementary DNA strands are chosen to define a fairly small target for amplification. These short primer sequences (usually around 20 nucleotides long) can be synthesized in the test tube. DNA carrying the sequences to be amplified is first melted at raised temperature (80–100 °C, depending on the base composition) to separate the two strands of the helix. Next the temperature is dropped sufficiently to allow the primers to find their matching sequences, and then adjusted again to allow the DNA polymerase to make the complementary strands (Fig. 3.3). When this temperature cycling is repeated (using an automated machine, a 'thermocycler') typically up to 30 times, this can result in the formation of microgram quantities of a defined stretch of DNA from a single molecule—an almost magical outcome! The trick here is the use of DNA polymerase enzymes that are not destroyed by the temperatures needed to melt the DNA helix at each cycle. Given that most enzymes are rendered inactive by high temperatures, where on earth do these come from? The answer is simple—those ancient thermophilic bacteria and archaea mentioned in Chap. 1! Applications of PCR are almost limitless (including in forensics and detection of biological contamination) as they allow sequencing and detection of particular target genes derived from un-purified DNA from 'dirty' sources. An important refinement of PCR has been the inclusion of indicators that allow the progress of the amplification to be monitored continuously ('real-time' or quantitative PCR—abbreviated to 'qPCR'). This allows the technique to be used to estimate the amount of target DNA originally present in the sample.

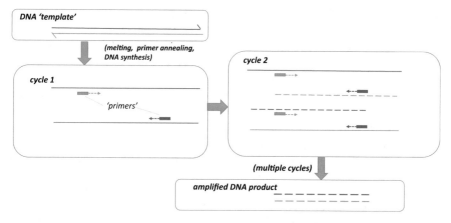

Fig. 3.3 Polymerase chain reaction (PCR). How specific sequences of DNA can be amplified in the test tube. The reaction temperature is repeatedly cycled, first to melt the two strands of the DNA helix, then to allow short primer sequences to pair (anneal) with the single DNA strands, and then to synthesise new complementary strands using a heat-tolerant DNA polymerase enzyme

The True Extent of Microbial Diversity

DNA sequence-based techniques have had particularly profound consequences for microbiological research, especially the study of microbial communities. As we have seen, microorganisms in nature are hardly ever found in isolation from other micro-organisms. Thus, although the 'pure culture' (i.e. one that is composed of a single strain of a single species) is an essential research tool that remains a gold standard of 'traditional' microbiological research, it hardly exists in nature. Any sample of soil, water or gut contents can contain hundreds of species per millilitre and can be considered a community. We know that there are certain genes that are common to all living cells, especially those essential genes concerned with the reproduction of DNA and the manufacture of proteins. For example, all living cells contain *ribo-somes*, which are tiny particles essential for the manufacture of new proteins. Ribosomes are composed of RNA molecules ('rRNAs') together with specific pro-teins. In bacteria, the genes that code for the universal 16S ribosomal RNA have proved extremely valuable targets for PCR amplification because they contain regions that are almost the same (i.e. they are highly conserved) in all bacterial species. These conserved regions are interspersed with other regions that are much more variable. The conserved regions provide convenient sequences for 'universal' PCR primers that allow amplification from multiple targets, whereas the sequences of the variable regions within the amplified product are characteristic of the partic-ular species of origin. What this means is that the amplified product (*amplicon*) from a complex mixture of bacteria (as found in a sample of soil or gut contents or faeces) will contain variable sequences representing all of the more abundant species within the sample. If we can just determine those sequences, then we have a description of the species composition of the original sample that has not required us to cultivate any of the organisms at all! The corresponding rRNA genes from Archaea and Eukaryotes require different primers, which makes it possible to do separate analyses of the diversity of these domains of life from the same samples (nb. the Eukaryote equivalent is '18S' rather than '16S', reflecting differences in their sedimentation characteristics by which they were first described).

This ability to perform culture-independent community analyses in this way has revolutionised the study of microbial ecology. At first it was still necessary to isolate the individual sequences by cloning them and applying the Sanger sequencing method, and the first analyses of the human gut microbiota were done in this way [6–8]. Subsequently another remarkable technical revolution, high-throughput sequencing (discussed later), removed the need for the cloning step, thus greatly speeding up the analysis.

By removing the requirement to grow micro-organisms in the laboratory, we remove a major element of 'cultivation bias' in describing the diversity of the microbial community. In many environments that are dominated by slow growing or nutritional fastidious organisms it has been estimated that as little as 1% of the true microbial diversity was being recovered by cultivation methods [9] and such cultivation-independent approaches have revealed the existence of major groups of micro-organisms that were previously unknown to science because of their failure to grow in the laboratory. We should also mention that culture-independent methods can introduce their own biases. For example, the ease of extraction of DNA varies widely between micro-organisms, and rates of amplification by PCR can also vary. For this reason, assessments of community diversity are best based on multiple approaches.

It should be clear from this that knowledge from culture-based studies and culture-independent analyses has to be pooled if we are to gain an overview of the true diversity of the whole microbial world. This is achieved by combining sequence information obtained from microbial isolates and from culture-independent methods into a *phylogenetic tree*. Branches on the tree correspond to *taxa*, by which we mean groups of organisms that are related through shared ancestry. By analogy with actual trees, we can regard members of the same 'twig' as the species, while those on the same small branch belong to the same genus, those joined via larger branches to the same family, and so on (Fig. 3.4). Those related at a still deeper level of branching belong to the same *phyla*.

All of the bacteria that we know about so far from culture-based studies are grouped into some 30 formally recognized phyla. In addition, however, there is a still higher number of 'Candidate Divisions' that have been revealed by culture-independent analysis alone and have no cultured representatives. These Candidate Divisions can go on to become phyla once cultured representatives have been isolated, so that the total number of bacterial phyla is potentially now at least 67 when the Candidate Divisions are included. The diversity of Archaea is still being uncovered. While relatively few phyla of cultured archaea have been described to date, there are many more potential phyla for which no cultured representative is available [10].

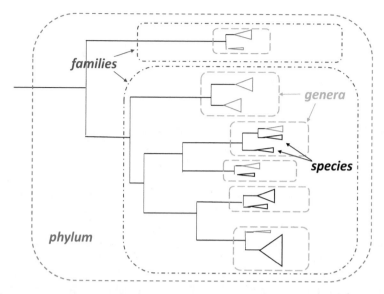

Fig. 3.4 A notional phylogenetic tree for a bacterial phylum derived from nucleotide or amino acid sequences. Horizontal lines represent the 'genetic distances' between branch points in the tree. Species, genera and families represent increasing degrees of sequence divergence. Triangles denote sequence divergence within species (i.e. at the strain level)

A bacterial or archaeal phylum is considered by taxonomists to be equivalent to the phyla that have been defined among multi-celled organisms. It is well worth noting that all animals with backbones, from snakes to dinosaurs to man, plus some less familiar sea creatures called Tunicates (sea squirts), are included in a **single** phylum, the 'Chordata'! This serves to emphasize the extraordinary diversity of the bacteria and archaea. Microbial Eukaryotes are similarly diverse, with 'Protists' belonging to 20 phyla and fungi to 6 Divisions (which are equivalent to phyla).

High-Throughput Sequencing and Meta- 'Omics' Approaches

The past 15 years have seen a massive change in DNA sequencing technologies to what is called "next generation" or high-throughput sequencing. These developments have dramatically decreased costs and greatly increased the speed with which data can be generated. Quite remarkably, the unit cost of sequencing DNA became 10,000 times cheaper in the 8 years leading up to 2015! One key advance was the ability to sequence multiple short DNA fragments in parallel using an automated *microarray*-based approach. The output consists of vast numbers of short sequences that (in contrast to Sanger sequencing) do not have a known start point. Instead, the approach relies on producing so many sequences that any one sequence must overlap with many others. This can be called 'shotgun sequencing'. Starting from the genomic DNA extracted from a single bacterial strain, for example, the resulting shotgun sequences are then matched and aligned on the computer to produce the most likely *consensus* sequences. While all the genes present in a genome can often be captured in this way, typically additional information derived from other methods is needed in order to produce a final continuous sequence that represents a complete genome. Most bacterial genomes turn out to exist within the cell as complete circles of DNA. While the genome is largely present in a single circular chromosome in most bacteria, this is often accompanied by smaller circles (called *plasmids*) that are capable of independent replication and in many cases of transmission to other cells in the microbial culture or community. Only exceptionally are there two circular chromosomes within a bacterial genome [11]. The first wave of high-throughput sequencing techniques are now in the process of being supplemented and indeed supplanted by others that employ fundamentally different technologies [12].

These new approaches to sequencing have had a massive impact on the study of microbial communities. First, they speed up the analysis of genomes from isolated (cultured) micro-organisms enormously, making it almost routine to compare hundreds of strains of a single species. This allows construction of what is called a '*pan-genome*' that identifies the core genes shared by members of the same species along with genes that vary between strains [13]. It has even become possible to produce a genome sequence from single bacterial cells. Secondly, high-throughput sequencing has greatly accelerated the sequencing of 16S rRNA gene amplicons produced from environmental samples by PCR and removes the need to clone individual sequences. This also removes biases that can occur in the cloning process. PCR primers used on different samples are 'signature-tagged' so that they can be sequenced simultaneously and the sequences then unravelled by computer analysis according to their origin.

Metagenomics In addition, however, these methods do something much more fundamental. Massive in-depth sequencing of DNA recovered from environmental samples has the potential to represent all the genes that are present in the more abundant strains and species found in that sample. The sequenced DNA potentially includes everything in the sample—viral, bacterial, archaeal and eukaryotic DNA sequences. Sequence analysis of DNA recovered from mixed microbial communities is referred to as *metagenomics* and represents a huge growth area for research.

Metagenomics raises some intriguing questions. How can you know which genes or DNA fragments within the massive datasets belong to which organisms, or indeed which DNA fragments have come from the same organism? Is it possible to reconstruct the genomes of individual organisms from such fragmentary information? Remarkably, these questions are steadily being resolved as the power and sophistication of computer sequence analysis (*bioinformatics*) grows. Some research groups are able to reconstruct what are referred to as 'metagenome species' from metagenomic sequence data, helped in part by the simple principle that sequences from the same organism should occur with the same frequency within the dataset.

Remarkably, it has even been possible to reconstruct microbial genomes from the highly complex rumen microbiome [14].

So, metagenomics makes it possible to catalogue all the genes present in a gut or faecal sample. In principle at least, it also makes it possible to reconstruct genomes for organisms that have not been cultured. Furthermore, metagenomic sequence information comes from the whole community, covering both bacterial and eukaryotic micro-organisms. Interpretation of such information however requires considerable care. For one thing, it is far from straightforward to infer function from sequences alone; a protein sequence predicted from a DNA sequence may match that of a known enzyme at 40% of its amino acid residues (a very significant level of *homology*), but this certainly does not guarantee that it has the same function. For another thing, DNA sequences obtained from meta-genomics provide information on potential, rather than actual, enzyme expression (e.g. a genome might code for 20 representatives of a given enzyme family, but only express one of them at any given time). On the other hand, there are other '-Omics' techniques that look at expressed mRNAs and proteins, respectively, which can tell us which genes are actually being transcribed and translated into products, and at what levels.

Transcriptomics and Proteomics Another important application of next generation sequencing is to analyse all the *RNA* molecules that are read off ('transcribed') from the DNA by the enzyme *RNA polymerase* within a cell population at any given moment. This powerful technique is known as *transcriptomics*. Production of messenger RNAs that are ready to be translated into protein sequences can differ markedly depending on the specific conditions of growth. This is because microbial cells control (regulate) transcription of different genes, so that their metabolism and activities ensure optimal growth and survival. Transcriptomics can tell us exactly which genes are switched on (expressed) and off in response to specific environmental triggers, for example leading to formation of the enzymes needed for utilization of a new growth substrate [15]. Just as genomic DNA sequencing can be applied to single cultures or to communities (metagenomics), so transcriptomic analysis can be performed on mixed microbial communities (*'meta-transcriptomics'*) as well as on single strains. Meta-transcriptomics has been applied very successfully to the rumen microbiota, taking advantage of the accessibility of this gut compartment for rapid sampling and processing that is required for analysis of microbial mRNA (see Chap. 4).

Sophisticated techniques are also available for detecting the proteins produced by microbial cells. These come under the collective term *proteomics*, or when applied to complex microbial communities *metaproteomics*. A clear advantage of this approach is that proteins carry out most gene functions within the cell. The proteins that are expressed in a given sample reflect translation of specific gene products (*mRNAs*) and are also the vehicles through which genes exert their effects. On the other hand, although detection techniques are advancing rapidly, it is as yet generally not possible to identify all, or even most, of the proteins present in a complex environmental sample.

Metabolomics The biochemical activities of microbial communities have important impacts on their environment. This is particularly true for microbial communities in the gut, whose products influence the nutrition and health of their host (Chap. 9). A range of new chemical techniques (referred to collectively as *metabolomics*) allow the measurement of almost all the chemicals of biological origin (metabolites) present in a given sample [16]. In humans and animals, the mix of metabolites found in samples of gut contents, faeces, urine and blood (the 'metabolome') includes products of the gut microbiota. Indeed, many metabolites detectable in human blood are microbial in origin. This means that it is possible to monitor the consequences of microbial activity via metabolomics without knowing anything about the microbiome itself! Metabolite profiles prove to be remarkably powerful means of detecting the impact of diet and of drug treatment [17].

What Do We Really Mean by and 'Microbiota Composition' and 'Gut Microbiome'?

Relative Versus Absolute Microbial Populations? Some important, albeit simple, points are often overlooked when discussing the gut microbiome. First, is the actual number of micro-organisms per gram of gut content important, or just the relative numbers of different micro-organisms? Am I healthier if I have 10^9 bifidobacteria per g gut contents, or if bifidobacteria make up 5% of my total gut bacteria? In practise, most sequence-based approaches only estimate relative numbers—i.e. the fraction or percentage of the total community that is made up of the various species, genera or phyla. So, if I have only 10^{10} total bacteria per g in my gut community, then 5% of bifidobacteria only gives me 5×10^8 of these bacteria per g. Another person might have 10^{11} bacteria per g, however, in which case 5% of bifidobacteria then corresponds to 5×10^9 bifidobacteria per g. As far as the sequence-based analysis is concerned, however, these two people would be identical. If we are concerned with beneficial metabolites (such as butyrate) then it is obvious that their production rates are more likely to relate to absolute rather than relative numbers of the relevant bacteria. Absolute numbers can be estimated by the technique of quantitative PCR (qPCR), although they are most accurately obtained by the use of fluorescent probes in microscopy or cell sorting.

On the other hand, it is also true that many health outcomes are likely to depend on a balance between beneficial and harmful elements of the microbiota. This is likely for example to apply to the effects of different elements of the gut microbiota that tend to either promote *inflammation* (pro-inflammatory) or 'damp down' inflammation (anti-inflammatory) via the host's immune system (see Chap. 10). Another very important balance is between pathogenic species that cause infection and commensal bacteria that help to provide a protective barrier against pathogens. For example, the ratio of potentially pathogenic enterobacteria against protective

Faecalibacterium prausnitizii in gut samples varies widely in different forms of colitis and may have diagnostic value [18].

The obvious conclusion is that, ideally, we need both absolute and relative numbers to properly assess the status of the gut microbiome. A neat compromise is to estimate total cell counts or total ribosomal RNA gene copies using a 'universal' detection method (DAPI staining of bacterial cells or qPCR with a broad bacterial primers). This then allows relative proportions based on sequencing to be converted approximately in absolute numbers for each subgroup. In the case of 16S rRNA gene-based data, however, we know that the number of gene copies varies between different groups of bacteria, which means that the relationship between copy number and cell number is not constant and has to be corrected for.

Which Community Do We Look At? A second major issue is, which gut microbial community are we measuring? Although *peristalsis*, resulting from constant muscle contractions in the gut wall, is responsible for mixing gut contents in the intestine, we know that there are significant differences in microbiota composition between different regions of the human intestine. This clearly applies longitudinally as we know that microbial numbers increase in the small intestine from the duodenum to the end of the ileum. The composition of the community also changes, from more facultatively anaerobic organisms and lactic acid bacteria at the start of the small intestine, to a more anaerobic community in the terminal ileum resembling that in the large intestine. We also need to consider that there are transverse and local varia-tions. The outermost of the two mucin layers that coats the large intestine has associated microbiota that probably differ at least quantitatively in species compo-sition from the digesta in the centre (lumen) of the colon. It is possible for surgeons to obtain biopsies of the gut lining during colonoscopic examination, although these yield small amounts of material and for obvious reasons their availability is subject to special circumstances, clinical priorities and patient consent. There is evidence that microbes associated with digesta particles in the gut lumen differ quantitatively from those in the liquid phase [19, 20].

It is clearly impractical to attempt to routinely assess all of these different microbial communities that make up the 'gut microbiome'. This is particularly true in healthy, free living, human volunteers. In practice, the great majority of work that is published on the human gut microbiome is based on the processing of stool (faecal) samples. It is easy to criticise the use of faecal samples, but they do offer some significant advantages. For one thing, a stool sample represents an accumulation of microbial cells that have passed through the gut over a period of time. Much microbial growth occurs at the start of the large intestine, in the caecum and *proximal colon* where the community first encounters undigested fibre, but it is expected that the resulting cells will move on to reach the rectum and so be represented in the stool sample. This assumption is broadly supported by observations.

Many technical issues have emerged around the storage and processing of faecal samples and variations in procedures and kits used in different laboratories. They undoubtedly mean that a 'faecal microbiota profile' cannot be taken as a universal or absolute measurement [21]. Freeze-thawing of un-processed stool samples can lead to the breakage of certain types of bacterial cell wall and differential losses of DNA from different groups of bacteria. In my own laboratory, we insisted initially on extracting DNA from freshly recovered faecal material in order to avoid such problems of frozen storage, although a compromise has now been developed that allows storage of samples in *cryo-protective* glycerol. We also know that some extraction kits fail to release DNA from microbial cells with particularly tough cell walls, while some 'universal' bacterial PCR primer sets targeting 16S rRNA genes may be quite biased in the sequences that they amplify. Nevertheless, faecal microbiota profiles have yielded, and continue to yield, extremely valuable insights into the behaviour of our gut microbiome.

Conclusions

We have emphasised here the massive impact that developments in DNA sequencing technology have had on microbial ecology. Most recent insights and discoveries on the role of gut microbiota in humans and animal have sprung from, or at least benefited from, the application of these technologies. Not surprisingly, these developments have had an equally profound impact on the biology of multicellular organisms. Two areas are particularly worth noting in relation to interactions between humans and their gut microbiota. First, the ability to compare the genomes of individual humans is producing entirely new information on inter-individual genetic variation, including its potential impact on the microbiota, as we will touch on in Chap. 10. Second, although it is largely outside the scope of this book, we should note that the ability to perform transcriptomics and proteomics on host cells and tissues can provide entirely new insights into interactions between host cells and gut micro-organisms.

References

 1. Barcenilla A et al (2000) Phylogenetic relationships of dominant butyrate-producing bacteria from the human gut. Appl Environ Microbiol 66:1654–1661
 2. Sijpesteijn KA (1948) Cellulose-decomposing bacteria from the rumen of cattle. PhD thesis, University of Leiden, Netherlands
 3. Chassard C et al (2010) The cellulose-degrading microbial community of the human gut varies according to the presence or absence of methanogens. FEMS Microbiol Ecol 74:205–213
 4. Derrien M et al (2008) The mucin degrader *Akkermansia muciniphila* is an abundant resident of the human intestinal tract. Appl Environ Microbiol 74:1646–1648
 5. Blattner FR et al (1997) The complete genome sequence of *Escherichia coli* K12. Science 277:1453–1462
 6. Suau A et al (1999) Direct analysis of genes encoding 16S rRNA from complex communities reveals many novel molecular species within the human gut. Appl Environ Microbiol 65:4799–4807

7. Hold GL et al (2002) Assessment of microbial diversity in human colonic samples by 16S rDNA sequence analysis. FEMS Microbiol Ecol 39:33–39
8. Eckburg PB et al (2005) Diversity of the human intestinal microbial flora. Science 308:1635–1638
9. Hugenholtz P et al (1998) Impact of culture-independent studies on the emerging phylogenetic view of bacterial diversity. J Bacteriol 180:4765–4774
10. Hugenholtz P (2002) Exploring prokaryotic diversity in the genomic era. Genome Biol 3: reviews0003.1
11. Wegmann U et al (2014) Complete genome of a new Firmicutes species belonging to the dominant human colonic microbiota ("*Ruminococcus bicirculans*") reveals two chromosomes and a selective capacity to utilize plant glucans. Environ Microbiol 16:2879–2890
12. Van Dijk EL et al (2014) The third revolution in sequencing technologies. Trends Genet 34:666–681
13. Rasko DA et al (2008) The pangenome structure of *Escherichia coli*: comparative genomic analysis of *E. coli* commensal and pathogenic isolates. J Bacteriol 190:6881–6893
14. Stewart PD et al (2018) Assembly of 913 microbial genomes from metagenomic sequencing of cow rumen. Nat Commun 9:art 870
15. Dodd D et al (2010) Transcriptomic analysis of xylan degradation by *Prevotella bryantii* and insights into energy acquisition by xylanolytic Bacteroidetes. J Biol Chem 285:30261–30273
16. Hollywood K et al (2006) Metabolomics: current technologies and future trends. Proteomics 6:4716–4723
17. Holmes E et al (2008) Metabolic phenotyping in health and disease. Cell 134:714–717
18. Lopez Siles M et al (2014) Mucosa-associated *Faecalibacterium prausnitzii* and *Escherichia coli* co-abundance can distinguish irritable bowel syndrome and inflammatory bowel disease phenotypes. Int J Med Microbiol 304:464–475
19. Walker AW et al (2008) The species composition of the human intestinal microbiota differs between particle-associated and liquid phase communities. Environ Microbiol 10:3275–3289
20. Swidsinski A et al (2008) Biostructure of fecal microbiota in healthy subjects and patients with chronic idiopathic diarrhea. Gastroenterology 135:568–579
21. Costea PI et al (2017) Towards standards for fecal sample processing in metegnomic studies. Nat Biotechnol 35:1069–1076

Chapter 4
How Microbes Gain Energy with and Without Oxygen

In most living organisms, sugars play the central role in supplying energy for cellular activities via common pathways that start from glucose. In heterotrophs, which include all animals and many micro-organisms, sugars come from carbohydrates that are taken up from the surroundings or consumed in the diet. Autotrophic green plants, algae and cyanobacteria make their own sugars through photosynthesis by using light energy, and then use them as sources of energy to fuel cellular activities.

Aerobic Respiration

The recovery of energy through the *oxidation* of sugars in *aerobic* organisms is referred to as *respiration*. Essentially, hydrogen is progressively removed from glucose and is finally combined with oxygen to form water, while the other end-product from glucose is 'hydrogen-free' carbon in the form of carbon dioxide. The hydrogens are removed by dehydrogenase *enzymes* and are passed on to *coenzyme* carriers (mainly the molecules NAD and NADP, derived from the vitamin nicotinic acid). With the addition of hydrogen these molecules are said to become *reduced*. The removal of hydrogen from glucose occurs in stages. First, a sequence of reactions converts glucose (which contains six carbons) into two molecules of the three-carbon compound, pyruvic acid, by the pathway known as *glycolysis*. Pyruvate is next converted into acetyl-coenzyme A (acetyl-CoA) with the loss of carbon dioxide. Coenzyme A, which is derived from pantothenate (vitamin B5), plays a key role along with thiamin (vitamin B1) and lipoic acid in facilitating this important enzyme reaction (Table 4.1). The two-carbon acetyl group is then released from the CoA to enter a cycle of reactions known as the 'Kreb's cycle' (after its discoverer) or as the 'TCA cycle' (for tricarboxylic acid) [1]. The TCA cycle results in further production of carbon dioxide and further removal of hydrogen via reduced coenzymes. Incidentally, acetyl-CoA also provides the entry point into the TCA cycle for the breakdown of fats.

© Springer Nature Switzerland AG 2020
H. J. Flint, *Why Gut Microbes Matter*, Fascinating Life Sciences,
https://doi.org/10.1007/978-3-030-43246-1_4

Table 4.1 Role of vitamins in microbial metabolism

Vitamin	Required for
Biotin	Activity of enzymes that add or remove carbon dioxide
Folic acid	Transfer of single carbon units
Hemin	Formation of cytochromes
Lipoic acid	Dehydrogenation of pyruvate
Nicotinic acid	Formation of NAD, NADP
Pantothenic acid	Formation of Coenzyme A
Pyridoxine (B6)	Formation of pyridoxal phosphate, required for certain enzyme reactions
Riboflavin (B2)	Formation of flavoproteins
Cyanocobalamin (B12)	Activity of mutase enzymes
Thiamin (B1)	Activity of multiple enzymes
Vitamin K	Formation of menaquinone
Para-aminobenzoic acid (PABA)	Transfer of single carbon units
Coenzyme M	Formation of methane by archaea

The key product of all this complex biochemistry is chemical energy, which is captured by production of the 'high energy' molecule *ATP* (adenosine tri-phosphate) from ADP (adenosine di-phosphate). ATP acts as the main energy currency unit of the cell that helps to drive (pay for) a whole array of chemical reactions that require energy input. These include the formation of crucial cell components such as proteins and nucleic acids.

ATP is formed during respiration in two fundamentally different ways. The ATP that is produced during glycolysis, and also from one of the steps in the TCA cycle, comes directly from enzyme reactions in which a phosphate is added from one of the substrates onto ADP ('*substrate-level phosphorylation*'). On the other hand, most of the ATP arising from aerobic respiration is produced by the enzyme ATPase, which is located in inner mitochondrial (or bacterial cell) membranes through the process known as '*electron transport phosphorylation*'. This is fuelled by the reduced coenzymes (mainly NADH) that are generated at various steps during the oxidative breakdown of glucose. The '*electron transport chain*' is also located in the inner mitochondrial (or bacterial) membrane and comprises an alternating arrangement of electron carriers (mainly *cytochromes*) and hydrogen carriers (coenzymes and flavoproteins).

A hydrogen atom consists of a proton and an electron. When the reduced coenzyme feeds hydrogen into the electron transport chain, this sequence of carriers leads to protons being picked up from one side of the membrane and ejected into the space between the two mitochondrial membranes (Fig. 4.1). The result is that protons are actively exported across the inner membrane, which otherwise does not allow them to pass through. At the end of the transport chain, two electrons are reunited with two protons and an oxygen atom to form water. Crucially, ATP is then formed by the membrane ATPase when the protons that have been exported re-enter

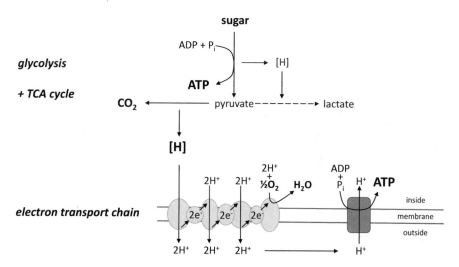

Fig. 4.1 The role of the mitochondrial electron transport chain in aerobic respiration. Hydrogen [H] and CO_2 are the main products of glycolysis and the TCA cycle, with some ATP formed directly. In the mitochondrion, protons (H+ ions) are released from hydrogen and electrons transported along the transport chain to be combined with oxygen to form water. The energy from the proton gradient is used for the formation of ATP from ADP (see text)

the cell, using the energy stored in the 'gradient' of proton concentration. This remarkable and now well-established mechanism was originally proposed by Peter Mitchell (who won the Nobel prize for this hypothesis) and is known as *chemiosmosis*. It is responsible for most of the energy that we humans (and indeed most living organisms) derive from our diets.

Aerobic respiration involving the TCA cycle yields far more energy per molecule of glucose than does glycolysis, but in contrast to glycolysis, it is critically dependent on a supply of oxygen. In mammals, the oxygen is supplied to cells from the bloodstream via the capillary network. If there is a deficiency of oxygen, glucose can only be converted as far as pyruvate which has then to be stored by conversion to another three-carbon molecule, lactic acid. In humans, accumulation of lactic acid is the cause of the muscle cramps that can set in following exertion and the lactate is only removed once oxygen becomes available again. Only two molecules of ATP are made by glycolysis from the conversion of a single molecule of glucose into lactic acid when oxygen is absent. By comparison, estimates made for the net ATP yield from the aerobic conversion of glucose to carbon dioxide and water are 26 in the bacterium *Escherichia coli* [2] or up to 32 in mitochondria. This is consistent with the very different energy changes predicted from chemistry for the two reactions (*free energy changes* of −47 kcal/mol for the formation of two lactate, compared with −686 kcal/mol for the formation of carbon dioxide and water, from glucose) [1].

Many micro-organisms that are tolerant of oxygen and able to grow in air have the ability to convert sugars such as glucose into carbon dioxide and water. Indeed, as noted earlier, the complex systems present in mitochondria that allow energy production to be linked to the utilization of oxygen first evolved in bacteria. Many other microbial groups, however, are not able to make use of oxygen and indeed many are incapable of growth in the presence of oxygen. These organisms are known as *anaerobes*. Some, such as the bacterium *Escherichia coli*, can switch readily between growth with and without oxygen and are known as *facultative anaerobes*. Those that cannot tolerate oxygen are known as *obligate anaerobes*. It should be clear as discussed above that in mammals the inability to use oxygen severely curtails the energy that can be gained from food. Obligately anaerobic micro-organisms have however evolved a wide variety of strategies for gaining energy from organic compounds.

Anaerobic Fermentation

Fermentation of sugars in the absence of oxygen means (by definition) that the products must show, collectively, the same ratio of hydrogen to oxygen as the starting material (the sugar). A number of solutions for meeting this apparently simple requirement have evolved over aeons of time among anaerobic micro-organisms. One simple solution that was mentioned already is the fermentation of one molecule of glucose to two molecules of lactic acid. This pattern is found in 'homo-lactic' fermenting bacteria, including many Lactobacilli and Streptococci. Since the elemental composition of lactic acid is simply half that of glucose there is no release of hydrogen or carbon dioxide. On the other hand, the predicted energy yield (at two ATP per glucose) is relatively small (Fig. 4.2a).

When brewer's yeast (*Saccharomyces cerevisiae*) is grown with little or no oxygen, it finds an alternative solution to the fermentation of sugars, which is to produce ethyl alcohol (ethanol). This simple fermentation is of course the basis of all of our brewing and wine-making industries. When used for baking, different strains of the same yeast species, growing in the presence of oxygen, convert sugars to carbon dioxide gas and water, thus making the bread rise.

As an alternative, in many micro-organisms pyruvate is converted to acetyl-CoA. Acetic acid (a two carbon acid) can then be formed from acetyl-CoA via the intermediate acetyl-phosphate, with the production of ATP. Production of acetyl-CoA from the three-carbon compound pyruvate entails the release either of carbon dioxide and hydrogen, or of formic acid. Since each glucose gives rise to two pyruvates, this means that a maximum of two acetates can be produced by this route, in theory resulting in four ATP (two from glycolysis and two from acetyl-phosphate) for each glucose fermented. There is however a biochemical snag here. The reduced NAD that results from glycolysis cannot release the hydrogen it carries unless the concentration of free hydrogen is extremely low. For many hydrogen-producing bacteria when growing alone, the only solution is to route this hydrogen into a *hydrogen sink* compound. The anaerobic rumen bacterium *Ruminococcus albus*, for example, was shown to produce ethanol and acetate from sugars when

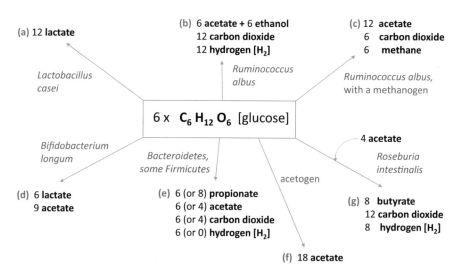

Fig. 4.2 Products of glucose fermentation by anaerobic bacteria. The simplest patterns are illustrated here (water not included). Fermentations often vary from those shown, with multiple products formed (acetate $C_2H_4O_2$; lactate $C_3H_6O_3$; propionate $C_3H_6O_2$; butyrate $C_4H_8O_2$; ethanol $C_2H_6O_1$; carbon dioxide C_1O_2; methane C_1H_4)

grown in isolation, with ethanol acting as the hydrogen sink (Fig. 4.2b). Because ethanol formation yields no ATP, this results in only two ATPs per glucose fermented, which is no better that the homo-lactic fermentation considered above. On the other hand, there are strong reasons to believe that this is not what actually happens within the gut microbial community. As was first demonstrated by Mike Wolin, rumen methanogens can remove the hydrogen very efficiently, producing methane [3]. The presence of a methanogen enables *R. albus* to produce two acetates and no ethanol from each glucose, with the formation of four ATP (Fig. 4.2c). Thus, the methanogenic archaeon provides an excellent partner that allows the hydrogen-producing bacterium to maximise its energy yield from glucose fermentation. This phenomenon is known as 'interspecies hydrogen transfer' and is considered to be a major factor in anaerobic fermentations in general, including fermentation in the gut. Many anaerobic bacteria release formate rather than hydrogen and carbon dioxide from pyruvate, and formate can also be used by methanogens to make methane.

Other hydrogen-utilizing ('hydrogenotrophic') organisms can convert hydrogen and carbon dioxide, or formate, into acetate. These bacteria, which are known as acetogens, use a special route called the Wood-Ljungdahl pathway and are thought to play a significant role in metabolite cross-feeding in the human colon (Fig. 4.3) [4, 5]. Interestingly some acetogens can convert glucose into three acetates by re-using the carbon dioxide (or formate) released from pyruvate to make the 'third' acetate (Fig. 4.2f). Since the Wood-Ljungdahl pathway requires an input of ATP, however, the extra energy gain is marginal. In the absence of sugars, some

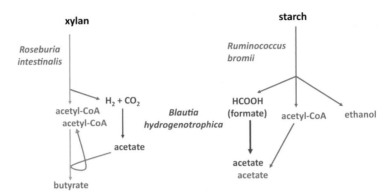

Fig. 4.3 Involvement of acetogenic bacteria in metabolic cross-feeding in the human colon. In these two examples, conversion of hydrogen and CO_2, or formate, to acetate by the acetogen (*Blautia hydrogenotrophica*) has been shown to enhance formation of butyrate and of acetate in co-cultures with *Roseburia intestinalis* [4] and *Ruminococcus bromii* [5]

acetogens can use hydrogen and carbon dioxide from the atmosphere to make acetate, although their growth is very slow.

Another distinctive fermentation is found in bifidobacteria. This 'bifid shunt' pathway is unusual in that it departs from the normal initial steps of glycolysis, but it can yield 1.5 molecules of acetate and one of lactate for each glucose fermented, with no release of hydrogen (Fig. 4.2d).

Two other 'short chain' fatty acids that are produced by significant subsets of gut anaerobes are propionate (a three carbon acid) and butyrate (a four carbon acid). Following the simple arithmetic of fermentation and based on their ratios of hydrogen to oxygen (6:2 for propionate, 8:2 for butyrate), it is clear that both of these compounds offer some utility as 'hydrogen sinks'. Propionate is typically produced along with acetate, in particular by bacteria of the Bacteroidetes phylum. If two molecules of propionate are produced for every one of acetate, then some carbon dioxide, but no hydrogen, is formed (Fig. 4.2e). In some species (or under some conditions of growth) succinate, the immediate precursor of propionate, rather than propionate is the end-product.

The production of butyrate is particularly intriguing. It is formed from two molecules of acetyl-CoA. Hydrogen and carbon dioxide (or formate) are of course formed from pyruvate along with the acetyl-CoA. Some of this hydrogen is however re-utilised in the steps leading up to the formation of butyrate (Fig. 4.2g). In addition, many butyrate-producing gut anaerobes, such as *Roseburia* species and *Faecalibacterium prausnitzii*, have the ability to import acetate, thus generating more acetyl-CoA and more butyrate and producing less hydrogen per glucose fermented (Fig. 4.4) [6].

There are of course many variations on these 'core' fermentation patterns that involve the production of several acids, or combinations of acids, alcohols and even solvents. In some species the fermentation products switch from acids to alcohols and solvents as the culture grows, presumably as a mechanism to avoid over-

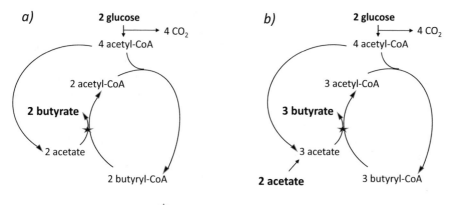

Fig. 4.4 Variable stoichiometry of butyrate formation. Many butyrate-producing bacteria that are dominant in the human intestine take up external acetate via the CoA-transferase reaction, increasing the yield of butyrate from fermentation. (**a**) shows no net acetate uptake, giving one mol butyrate per glucose; (**b**) shows net acetate uptake, in this case giving 1.5 mol butyrate per mol glucose fermented. Acetate uptake and hence butyrate yield increases with lower pH [6]. (nb. Net hydrogen production also changes, but is not shown here)

acidification. One of the first such fermentations to be discovered, which turned out to have more than simply scientific significance, was the production of acetone and butanol by *Clostridium acetobutylicum*. The fact that acetone could be used in the manufacture of cordite for munitions, and that the discovery was made at the time of the First World War, led its discoverer, the organic chemist Chaim Weizmann, to make an agreement for use of his fermentation process to supply the British military with microbially-produced acetone. Many years later, Weizmann became the first President of Israel.

Thus far we have assumed that, under anaerobic conditions, ATP is only produced directly from enzyme reactions either from glycolysis or from a few other substrate-level phosphorylation reactions such as the formation of acetate from acetyl-phosphate. On the other hand, some anaerobes are able to use compounds such as sulfate or nitrate in roles similar to that played by oxygen in aerobic micro-organisms. This process is known as *anaerobic respiration* and is carried out by electron transport chains similar to those of aerobic bacteria, resulting in ATP formation by ATPases, driven by proton gradients. These inorganic (non-carbon based) molecules act as hydrogen-acceptors (i.e. they become reduced) in the same way that oxygen is reduced to water. Thus sulfate-reducing bacteria (SRB) reduce sulfate to sulfide by means of anaerobic respiration, while other micro-organisms reduce nitrate to ammonia.

Most fermentative anaerobes lack an electron transport chain and are unable to perform anaerobic respiration. Nevertheless, it is becoming clear that even in these organisms, ATP formation by ATPases can often be driven through little-studied chemiosmotic mechanisms by gradients of protons, or possibly other *ions* such as

Na$^+$. This is highly significant as it means that the energy yield from fermentation is likely to be substantially higher than is predicted from substrate-level ATP formation alone. Anaerobic bacteria and archaea have had longer than any other life forms to evolve efficient mechanisms for energy production, but we are only now starting to understand the biochemical tricks that allow them to survive on the planet.

As we will see later, these different solutions to the balance of fermentation are not just of interest to a small cadre of biochemists and anaerobic microbiologist 'nerds'. Changes in the gut environment or food supply together with shifts in the composition of the microbial community can alter the balance of acids and gases produced by the gut microbial community. These have real significance and impact both for human health and for the environment.

Metabolite Cross-Feeding

Inter-species hydrogen transfer, discussed above, is just one example of the broader phenomenon of metabolite cross-feeding, which is a fundamental feature of microbial communities. Apart from formate, several other fermentation acids that are major end products of pure bacterial cultures are normally detected only at low levels in the mixed microbial community. This is generally true for lactate, which can be converted to butyrate, propionate or acetate by other members of gut microbial communities (Fig. 4.5). The same is also true for succinate, which is generally converted efficiently to propionate. Accumulation of lactate and succinate in gut samples can be an indicator of altered microbiota composition and function. Another form of cross-feeding, involving the interchange of partial degradation products from complex substrates, will be discussed in Chap. 7.

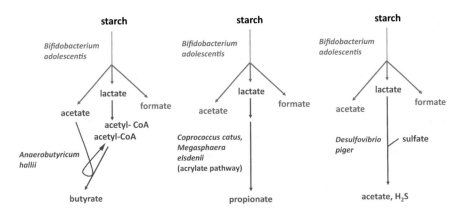

Fig. 4.5 Alternative routes for metabolic cross-feeding of lactate in anaerobic microbial communities. Lactate can be produced by many gut anaerobes from a variety of substrates, but is exemplified here with starch fermentation by *Bifidobacterium adolescentis*

More fundamentally, many micro-organisms do not have the ability to make all the molecules that they require for growth and reproduction. Instead they rely on obtaining factors needed for growth via cross-feeding from other organisms that are able to make them. Two of the complex cell constituents essential to life are proteins and nucleic acids. While all living cells have the capacity to make these molecules, not all of them make all the building blocks of which they are composed. Proteins consist of folded chains of amino acids that are joined together within the cell in precise sequences as dictated by the cell's genetic material. Proteins not only provide the enzymes that, as *catalysts*, drive the diverse chemical reactions necessary for life, but also many structural components of the cell. Twenty different amino acids are required to make proteins, but it is quite common for a micro-organism to be unable to make enough of each of these amino acids for its own needs. The deficiency is generally made up from the cell's immediate environment, by taking up amino acids that have been made by other organisms. The same is true for nucleic acids, which fulfil the vital role of encoding and transcribing the cell's genetic information. Not all microorganisms can make the four *pyrimidine* and *purine* bases that are found in all nucleic acids and again many rely on cross-feeding. Another important group of 'growth factors' that frequently have to be supplied by cross-feeding are the vitamins, a diverse group of complex molecules that are needed to help drive particular enzyme-catalyzed reactions (Table 4.1).

Some chemical conversions cannot be performed by a single organism alone because they are energetically unfavourable (i.e. they have a positive free energy change). The presence of a second organism that removes a reaction product can however make the combined reaction favourable, as we saw earlier in the example of interspecies hydrogen transfer involving *Ruminococcus albus* and methanogenic archaea. This type of interaction is known as *syntrophy* [7].

Why Strict Anaerobes?

Before we move on, there is an obvious question—why do obligate (or 'strict') anaerobes exist at all? Why has evolution and natural selection not resulted in all organisms being aerobic, or at least facultatively anaerobic like *E. coli*, given the evident advantages of using oxygen to maximise energy gain from food sources? There are several answers to this question. First, as noted already, oxygen was absent or present at extremely low concentrations in the earth's atmosphere for the first 3 billion years of the earth's history during which time the first microbial life forms evolved. These first life forms were therefore all initially anaerobic, gaining energy through processes that did not require oxygen (although photosynthesis by cyanobacteria of course produced the oxygen that progressively changed the atmosphere). For a variety of reasons, cellular machinery that first evolved to function in the absence of oxygen will have needed major modification to function when oxygen became a factor in the environment. In some cases this appears to have been impossible. Nitrogen fixation in cyanobacteria has to occur in special structures known as heterocysts that afford protection from oxygen because the nitrogenase enzyme that carries out nitrogen fixation is damaged by oxygen. This is in spite of the fact that the other cells within the same filamentous chain are photosynthetic and not only are tolerant of oxygen but are busy producing it. More generally, oxygen and its reactive products (or 'radicals') inactivate certain enzymes (especially those carrying iron-sulfur clusters) that play a central role in the fermentation pathways of anaerobic bacteria. Species that tolerate oxygen have evolved protective mechanisms that help to avoid damage to their cellular components. Specifically, the enzymes superoxide dismutase (SOD) and catalase dispose of the superoxide radical and of hydrogen peroxide, respectively. Even so growth may be still be inhibited by oxygen in species that possess these enzymes [8].

Anaerobic micro-organisms show a very wide range of oxygen tolerance. The most strict anaerobes (such as methanogenic archaea) fail to grow in all but the most reduced environments and are killed by even a trace of oxygen. Other obligate anaerobes (such as human colonic *Bacteroides*) can survive, although they cannot grow, when exposed to air for several hours but have the ability to grow again when anaerobic conditions are restored. Meanwhile some anaerobes that are killed by exposure to air, are nevertheless able to make use of oxygen to assist growth when it is present at very low concentrations [9]. Terms such as 'nan-aerobe', or 'micro-aerophile' have been used to describe behaviour in response to low oxygen concentrations.

While oxygen is now 20% of atmospheric gas, very many present-day habitats are deficient in it, including for example waterlogged soils, sediments, swamps, deep seas, and these create anaerobic environments that favour specialist micro-organisms. Many of the present-day anaerobes that occupy these environments retain features of the early colonisers of the planet that existed before the rise in atmospheric oxygen. In situations where the supply of oxygen is limited, we know that facultatively anaerobic organisms tend to grow quickly, but then rapidly use up the available oxygen. This can create conditions where organic carbon remains plentiful, but oxygen concentrations are very low. Such conditions are then tailor-made for exploitation by anaerobic organisms that have the ability to make use of organic carbon, particularly those forms of carbon (such as cellulose) that are less readily accessible to other micro-organisms. This is exactly what happens in the most densely populated communities within the animal gut, such as the rumen and large intestine of mammals, that are dominated by obligate anaerobes whose numbers can reach 10^{11} (one hundred billion) cells per millilitre. In the absence of oxygen, obligate anaerobes clearly outcompete facultative anaerobes. It is not immediately obvious why this should be so, but we must conclude that obligate anaerobes are more efficient at gaining energy from available substrates and at tolerating the prevailing environmental conditions in the absence of oxygen. A key factor may be that such dense anaerobic microbial communities create their own environments by producing high concentrations of fermentation acids.

By no means all present-day anaerobes can be considered as 'relics' from the early evolution of life on earth. For some there is evidence that the anaerobic state is a secondary characteristic that became selected in previously aerobic ancestors. The anaerobic fungi of the rumen provide a neat example. Until the 1970s certain structures visible in rumen contents under the microscope were assumed to be spores of 'protozoa' and all fungi were assumed to be aerobic. The discovery that these structures were in fact fungi that were strictly anaerobic, and even lacked mitochondria, was a complete surprise to researchers at the time. It is thought that these organisms evolved from aero-tolerant 'chytrid' fungi that were present in plant material ingested by herbivorous animals, subsequently losing the ability to use oxygen under the highly reduced (anaerobic) conditions of the rumen. This can be compared to the loss of vision in some animal species that inhabit dark caves and is a striking example of the specialism that can accompany successful colonisation of a given habitat by micro-organisms. Anaerobic fungi are found in other herbivorous mammals such as horses, although these have not been reported from the human gut where transit times are faster.

The Rumen Fermentation

The rumen provides an excellent model for anaerobic gut metabolism that has been studied over many decades. As we will see, its behaviour and manipulation have great significance not only for agricultural production and animal welfare, but also for environmental sustainability.

The three domains of life are all represented within the rumen microbiota. Most numerous are the bacteria, especially two phyla, the Bacteroidetes and the Firmicutes, that normally dominate the community, but significant numbers of Fibrobacteres, Actinobacteria and Proteobacteria are also present. The archaeal population comprises mainly methanogens that convert formate or carbon dioxide and hydrogen into methane. Rumen microbial eukaryotes include both anaerobic protists and anaerobic fungi. The protists are less numerous but generally much larger than the bacteria and are estimated to account for 50% of the microbial biomass in the rumen. Most of these fascinating organisms are 'ciliates' that carry a mass of tiny hairs (cilia) on their cell surfaces that enable them to swim. They have the ability to engulf particles of food, which can include both plant material from the diet and bacterial cells. Indeed, the rumen protists act as predators upon the bacterial population, increasing the turnover of bacterial cells. The anaerobic fungi are an unusual and highly specialised group whose filamentous (hyphal) growth habit helps them to invade and break down plant fibre. As noted earlier, they are thought to have evolved from aerotolerant chytrid fungi associated with plant tissues.

The highly efficient breakdown of plant fibre in the rumen involves the combined activities of specialized bacteria, fungi and protists [10]. Only a few species of rumen bacteria have the ability to attack plant cell walls by themselves. These include certain *Ruminococcus* species and *Fibrobacter succinogenes* that produce a wide range of enzymes active against cellulose and hemicellulose (the main constituents of the plant cell wall). The anaerobic fungi produce a similarly wide range of these enzymes. In the bacterium *Ruminococcus flavefaciens* and in rumen fungi, the relevant enzymes are organised on the outside surface of the cell into complex assemblies known as *cellulosomes*. Several of the ciliated protists have also been shown to produce their own cellulases and hemicellulases and are assumed to degrade plant fragments after engulfing them within food vacuoles.

This complex rumen community produces a mixture of mainly organic acids and gases from the fermentation of plant material. The proportions of these fermentation products are not fixed, however, and their exact balance has important consequences for animal health, animal production and the environment. The absorption of the short chain (volatile) fatty acids acetate, propionate and butyrate can provide the animal with up to 80% of all the calories that it gains from its plant-based diet. Without this energy from microbial fermentation the animal simply could not survive on the non-digestible plant fibre that makes up the bulk of its natural diet.

The fermentative activities of the rumen bacteria and microbial eukaryotes also result in much carbon being released in gaseous form as carbon dioxide. Since all ruminants harbour methanogenic archaea, a fraction of this of this CO_2 is combined with hydrogen to produce methane (CH_4). Both methane and CO_2 are 'greenhouse gases' whose increasing concentrations in the atmosphere are contributing to global warming. Methane is of particular concern however as it is some 30 times as potent as CO_2 as a greenhouse gas. Remarkably, agricultural livestock are the largest source of methane on the planet that can be attributed to human activity. As a result, attempts to understand and control methane emissions from ruminants have been a major focus of recent research.

The production of methane by ruminants is influenced by the diet, with poorer diets that contain more indigestible fibre yielding more methane than diets made up of more readily digestible concentrate feed. It has been known for some time that there is a link between methane production and agricultural productivity. Methane is considered by animal nutritionists to represent a loss of carbon that would otherwise be available to the animal if only it could be re-routed into the fatty acids that provide sources of energy. Various approaches that involve inhibiting methane formation have been explored over the years, some of them successfully, with the intended outcome being to re-route hydrogen from methane formation into 'hydrogen-sink' acids such as propionate and butyrate.

In addition, methane production is apparently affected by the animal itself [11]. Some fascinating recent work has indicated that animals within a herd that show the highest production efficiencies tend to produce less methane than those with the lowest production efficiencies. What is more, low methane formation correlates with the species composition of their rumen microbiota, both in cows [12] and in sheep [13]. In particular, the low methane emitting animals tend to show greater populations of bacteria that produce and consume lactic acid. Lactate produced by the rapid fermentation of sugars in these animals is efficiently converted by other bacteria mainly into propionate or butyrate. It is argued that this results in less net hydrogen formation than the direct fermentation of sugars to produce the acids, thus limiting methane formation.

The reasons for such variation in the rumen microbiota between animals, which appear to be at least in part heritable, have yet to be fully pinned down. One plausible suggestion is that low methane production occurs in animals with smaller rumen sizes and more rapid transit of contents through the rumen. This is likely to influence the composition and metabolism of the rumen microbiota through effects on rates of fermentation and rumen pH [13]. The supply of readily fermentable feed together with rapid rumen transit may tend to curtail the growth of slowly growing hydrogen-producing organisms and methanogenic archaea, allowing the racier 'sport cars' in the form of faster growing lactate-producing bacteria to take over.

Although they are not ruminants, kangaroos and wallabies are herbivores that also depend on fermentation in their foregut. Interestingly, this fermentation releases considerably less methane than does rumen fermentation in domestic livestock. It is suggested that this may be due to high abundance of gut bacteria that produce succinate, which provides another 'sink' for hydrogen [14].

The formation of too much lactic acid can have very negative consequences for the ruminant. *Lactic acidosis* can result from supplying excess starchy concentrate feed and can result in death of the animal. The supply of readily available starch promotes the growth of lactic acid producing bacteria such as *Streptococcus bovis*. As lactic acid production exceeds the ability of lactate-utilizing bacteria within the community to remove it, this results in an accumulation of lactate and a decrease in rumen pH (lactic acid is significantly more acidic than acetate, propionate and butyrate). Lower pH values below 6 in turn inhibit the growth of most of the rumen organisms that dominate the community when the pH is in the normal range between 6 and 7. Bacteria that are tolerant of lactic acid, such as streptococci and lactobacilli, come to dominate the community at lower pH. Some of these produce the D-form of lactate, which is a neurotoxin that contributes to the acidosis in the animal. If not reversed, lactic acidosis can be fatal.

It should be apparent that lactic acidosis is at the extreme end of a continuum that leads from animals with low production efficiency through to highly efficient production (Table 4.2). Across this range, mainly in response to the type of diet supplied, rumen pH decreases and the fibrolytic microbiota normally characteristic of extensively grazing animals is steadily replaced by faster growing organisms.

Table 4.2 States of the rumen fermentation

Husbandry	Extensive → Intensive			
Diets	Grazing animals, high intakes of plant fibre	Intensive rearing, concentrates in feed. Faster transit?	High concentrate intakes	
Rumen community	Fibre-degrading community; hydrogen producers, methanogens	Favours starch-degrading bacteria, lactate-utilizers	Shift towards lactic acid bacteria, proteobacteria	Lactic acid bacteria dominant
Fermentation	Hydrogen production and methanogenesis	Less hydrogen, more propionate, butyrate, less methane	Sub-acute rumen acidosis	Lactic acidosis
Rumen pH	7 (neutral)	6 (slightly acidic)	5 (acidic)	4

Also within this range are conditions known as subacute rumen acidosis that are less extreme than lactic acidosis, but nonetheless damaging to animal health.

The rumen fermentation has provided many important insights into the behaviour of anaerobic microbial communities that have parallels in other anaerobic gut communities. These include the human large intestine, to which we will turn our attention in the chapters that follow.

Conclusions

Anaerobic microbial communities similar to that of the rumen are found in the hind gut of most herbivorous mammals. The horse large intestine harbours large numbers of anaerobic bacteria and fungi that actively ferment cellulosic plant material to short chain fatty acids. The problem of lactic acidosis is also encountered in horses. As we have seen, oxygen is a key factor controlling microbial growth and metabolism and its concentrations vary across and along the mammalian gut. Even in the large intestine and rumen, the walls of the gut are well supplied with oxygenated blood and oxygen diffuses into the gut itself. This is likely to create special microenvironments or niches in which the less oxygen-sensitive organisms will thrive. In other parts of the gastrointestinal tract, notably the small intestine, flow rates and oxygen concentrations are higher and microbial numbers are lower than in the large intestine. As a result, the species composition and activities of these communities are very different from those of the rumen and large intestine.

References

1. Lehninger AL (1979) Biochemistry, 2nd edn. Worth Publishers, New York
2. Gottschalk G (1979) Bacterial metabolism. In: Starr MP (ed) Microbiology. Springer, Berlin
3. Wolin M et al (1997) Microbe-microbe interactions. In: Hobson PN, Stewart CS (eds) The rumen microbial ecosystem. Springer, London, pp 467–491
4. Chassard C, Bernalier-Donadille A (2006) H_2 and acetate transfers during xylan fermentation between a butyrate-producing xylanolytic species and hydrogenotrophic microorganisms from the human gut. FEMS Microbiol Lett 254:116–122
5. Gomez JL et al (2019) Formate cross-feeding and cooperative metabolic interactions revealed by transcriptomics in cultures of acetogenic and amylolytic human colonic bacteria. Environ Microbiol 21:259–271
6. Louis P, Flint HJ (2017) Formation of propionate and butyrate by the human colonic microbiota. Environ Microbiol 19:29–41
7. McInerney MJ et al (2008) Physiology, ecology, phylogeny, and genomics of microorganisms capable of syntrophic metabolism. Ann NY Acad Sci 1125:58–72
8. Pan N, Imlay JA (2001) How does oxygen inhibit central metabolism in the obligate anaerobe *Bacteroides thetaiotaomicron*. Mol Microbiol 39:1562–1571
9. Khan MT et al (2012) The gut anaerobe *Faecalibacterium prausnitzii* uses an extracellular electron shuttle to grow at oxic-anoxic interphases. ISME J 6:1578–1585

10. White BA et al (2014) Biomass utilization by gut microbiomes. Annu Rev Microbiol 68:279–296
11. Roche R et al (2016) Bovine host genetic variation influences rumen microbial methane production with best selection criterion for low methane emitting and efficiently feed converting hosts based on metagenome gene abundance. PLoS Genet 12:e1005846
12. Ben Shabat SK et al (2016) Specific microbiome-dependent mechanisms underlie the energy harvest efficiency of ruminants. ISME J 10:2958–2972
13. Kamke J et al (2016) Rumen metagenome and metatranscriptome analyses of low methane yield sheep reveals a *Sharpea*-enriched microbiome characterised by lactic acid formation and utilisation. Microbiome 4:56
14. Pope PB et al (2011) Isolation of Succinovibrionaceae implicated in low methane emissions from Tammar Wallabies. Science 333:646–648

Chapter 5
Who Inhabits Our Gut? Introducing the Human Gut Microbiota

While all regions of our digestive tract are colonized by micro-organisms to some degree, the densities, composition and nature of the associated microbial communities change markedly as we travel down through the gut (Fig. 5.1). The mouth (oral cavity) is well colonised, particularly the surfaces of teeth and crevices between the tooth and gum. As semi-permanent features, teeth allow the development of microbial biofilms, in which different species attach to each other and to the tooth surface in a highly structured manner [1]. Microbial biofilms can develop elsewhere, for example on implanted catheters, but the rapid turnover of epithelial cells and digesta particles probably limits their formation further down the gut. Infections of teeth and gums can have a major impact on health not only through tooth decay, but also through their effects on the heart, brain and immune system. These problems are due to pathogenic oral bacteria such as *Streptococcus mutans* and the anaerobe *Porphyromonas gingivalis*, a member of the Bacteroidetes phylum. The protein-destroying enzyme gingipain, which is produced by *P. gingivalis* was recently proposed as a factor in the development of Alzheimer's disease [2]. Rapid passage of gut contents after swallowing means that the healthy oesophagus shows very low numbers of micro-organisms, while the highly acidic stomach also deters most colonisers. Some specialist microorganisms do however manage to colonise the stomach wall, notably *Helicobacter pylori* [3].

Food leaves the stomach to enter the duodenum, followed by the jejunum and then the ileum. These different parts of the small intestine include the main sites for digestion of food by enzymes such as trypsin (protein breakdown) and amylase (starch breakdown), which are released from the pancreas into the duodenum, and for absorption into the bloodstream of the nutrients released by digestion. The walls of the small intestine carry huge numbers of finger-like protruberances called *villi* that create a massive surface area for nutrient absorption. Rapid transit of gut contents in the small intestine limits microbial colonisation, but microbial cell densities increase steadily along its length to reach relatively high values (100 million (10^8) cells/ml) in the ileum. The walls of small intestine, particularly the ileum, also contain large numbers of cells that form part of the immune system (GALT or 'gut-

© Springer Nature Switzerland AG 2020
H. J. Flint, *Why Gut Microbes Matter*, Fascinating Life Sciences,
https://doi.org/10.1007/978-3-030-43246-1_5

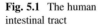 **Fig. 5.1** The human intestinal tract

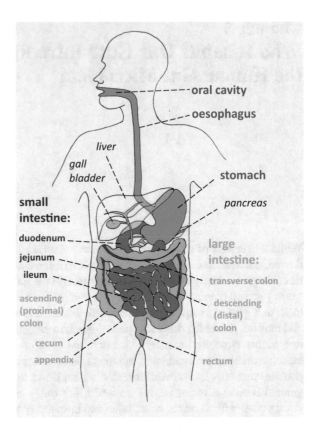

associated lymphoid tissue'). This is therefore a prime site for microbial interactions with the immune system, including immune sampling (Chap. 10).

The large intestine harbours the highest microbial cell densities, which can be up to 100 billion (10^{11}) cells per gram of gut contents. This high density reflects the much slower *transit* compared with the small intestine, allowing microbial growth to be driven by the fermentation of fibre sources. Together with the large volume of this compartment, this means that the large intestinal microbiota account for the bulk of our gut microbiota overall. In contrast to the small intestine, the healthy colon is lined with thick layers of protective mucus that separate the gut microbiota from the epithelial cells that line the colon wall. There are two layers of this mucus. The innermost one (closest to the colon wall) is very dense and essentially free of micro-organisms, while the outer layer is more diffuse and does contain some micro-organisms, including some that gain energy by degrading mucus. The localized breakdown of this mucus barrier, which then results in closer contact between gut bacteria and host cells, is one of the problems associated with inflammatory bowel disease.

The human large intestine starts with the short region known as the caecum, to which is attached a very short sac, the appendix. The appendix is the site of troublesome bacterial infections and its surgical removal (once done as a preventative measure) is rather common. The caecum is continuous with the ascending (or right-sided) colon which leads to the transverse colon, and then to the descending (left-sided) colon and rectum. Most of our information on the gut microbiota comes from analysis of faecal samples, which are assumed to reflect mainly the microbiota of the descending colon. In fact, the few studies that have been conducted indicate there may be little difference between the microbiota composition at different sites along the length of the colon, and even by comparison with the terminal ileum [4]. The microbiota composition of higher regions of the small intestine, the duodenum and jejunum, are however very different, with higher proportions of facultative anaerobes. Data are decidedly scarce since these are the most difficult of all regions of the gut to sample in a healthy individual.

The microbial community of the human colon is generally dominated by obligately anaerobic bacteria. As in the rumen, any available oxygen in the colonic contents is quickly used up by facultative anaerobes, including Proteobacteria and lactic acid bacteria, leading to conditions in which obligate anaerobes prevail. The active anaerobic fermentation of undigested fibre by the resident microbial community of the ascending colon raises the concentrations of short chain fatty acids. This leads to a drop in pH of up to 3 units at the start of the colon compared with that in the terminal ileum, making the colon slightly acidic. This pH response to fermentation of fibre has been documented in healthy humans using *telemetry* [5].

Bacteria in the Human Large Intestine

Bacteria are the most numerous inhabitants of the healthy human intestine. Almost all of the bacteria found in human gut belong to just five or six phyla, two of which are generally predominant in the healthy adult large intestine. Representatives of the most abundant bacteria identified in faecal samples from human adults have mostly been cultured, although some are still waiting to be named [6, 7] (Table 5.1; Fig. 5.2). There is little support for the widely stated opinion that most human gut bacteria are 'unculturable'. In marked contrast to the microbiomes of oceans and sediments, where microbial growth rates are generally very slow, the gut community exists in an open tube and its members simply get 'washed out' unless they can grow fast enough to keep pace with the constant through-flow of gut contents. We are now coming around to the view that most gut bacteria can be cultured, given the right media and growth conditions, it just takes considerable time and effort to achieve this. As a result, many of the less common microbial inhabitants of our gut have yet to be studied in culture. Also, since most effort has been directed thus far at urbanised human populations in Europe, Japan and the USA, even the dominant micro-organisms from many other parts of the world may not have been recovered as cultures. Since it is far easier to profile communities from faecal samples by culture-independent (molecular) methods (Chap. 3), perhaps this does not matter—why should we bother with culturing? The answer, which will be expanded later, is that

Table 5.1 Some important bacterial species within the human intestinal microbiota

phylum: **Bacteroidetes**		*phylum:* **Actinobacteria**
Bacteroides vulgatus *Bacteroides uniformis* ***Bacteroides thetaiotaomicron*** *Prevotella copri* *Bacteroides fragilis*	*phylum:* **Firmicutes** *Eubacterium rectale* *Roseburia* **spp.** ***Anaerobutyricum hallii*** *Anaerostipes hadrus* ***Blautia* spp.**	***Bifidobacterium adolescentis*** ***Bifidobacterium longum*** *Colinsella aerofaciens*
phylum: **Proteobacteria**	*Coprococcus catus* ***Faecalibacterium prausnitzii*** *Ruminococcus bromii*	*phylum:* **Verrucomicrobia**
Sutturella *Desulfovibrio piger* *Bilophila wadsworthii* *Escherichia coli* *Shigella* *Salmonella enterica* *Helicobacter pylori* *Campylobacter jejuni*	*Dorea longicatena* *Veillonella parvula* *Clostridium ramosum* ***Lactobacillus casei*** *Clostridium difficile* *Enterococcus faecalis*	***Akkermansia muciniphila*** *phylum:* **Fusobacteria** *Fusobacterium nucleatum*

Species consisting mainly of commensal strains are shown in blue (bold indicates proposed or actual probiotic/therapeutic use). Species with many pathogenic strains are show in red. Purple indicates harmful activities. See text for details

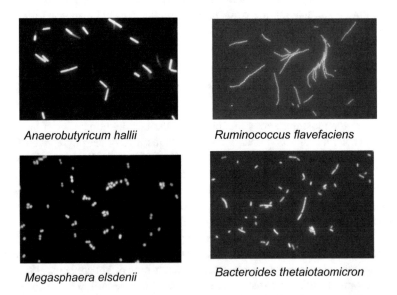

Anaerobutyricum hallii   *Ruminococcus flavefaciens*

Megasphaera elsdenii   *Bacteroides thetaiotaomicron*

Fig. 5.2 Gut bacteria in pure culture visualised by fluorescent in situ hybridisation (FISH) probing. *A. hallii*, *R. flavefaciens* and *M. elsdenii* are species of Firmicutes; *B. thetaiotaomicron* belongs to the Bacteroidetes. Micrographs are from Dr Alan Walker (PhD thesis, U Aberdeen) with permission

cultured isolates provide crucial information that is impossible to gain simply from DNA sequences.

Proteobacteria

Proteobacteria is a phylum of 'Gram-negative' bacteria[1] that includes many bacteria that we know as pathogens, such as *Escherichia coli, Salmonella typhimurium* and *Shigella dysenteriae* among others. Gram-negative cells are bounded by two distinct membranes composed of phospholipids and proteins that sandwich a zone called the *periplasm*. Also characteristic of these bacteria is an outermost layer composed of molecules called lipopolysaccharides (LPS). The human immune system reacts strongly to LPS as an antigen, providing it with an early warning signal of infection by Gram-negative pathogens. These pathogens can cause damage to the host in a variety of ways. Some actively secrete *toxins* that interfere with cellular processes in the host and can cause severe tissue damage. A common group of diarrhoea-causing *E. coli* (Enterotoxigenic or 'ETEC' strains) produce both a heat-labile toxin that is inactivated by cooking, and a heat-stable toxin. Heat-stable toxins are not destroyed by cooking, even though the cells that produce them are killed, meaning that toxin-producing strains can cause damage to the host by being present in food without even growing in the gut. Not all pathogenic strains of *E. coli* produce toxins, but those that do not can still cause damage to the surface layers of the gut (the gut *epithelium*) or can have the ability to cross over the barrier provided by the gut wall and to invade host cells. The notorious pathogenic *E. coli* 0157 strains produce a damaging 'Shiga-like' toxin and are also capable of tissue invasion.

We know more about *Escherichia coli* than we do about any other bacterium. In common with many gut Proteobacteria, *E. coli* is tolerant of oxygen, although it is also able to grow in its absence, which makes it a facultative anaerobe. This feature aided its early isolation in 1886 by the German paediatrician Theodor Escherich from pus on bandages. Apart from its considerable medical importance, *E. coli* has become an invaluable research 'work horse' organism for studies on the fundamental mechanisms of inheritance and cellular biochemistry. Despite its well-earned reputation as a pathogen, many strains of *E. coli* that inhabit the gut lack the sets of *virulence genes* that characterise pathogens and, as a result, do no harm to their host. These are harmless 'commensal' strains and for obvious reasons they are the ones preferred for the purposes of basic laboratory research work!

The distinction between pathogen and commensal is not always clear-cut. A group of *E. coli* (adherent invasive, 'AIEC') that have only recently been recognized as pathogens appear to contribute to the pathology of Crohn's disease under conditions where the integrity of the gut barrier is disturbed but are not known to cause the same problems in other individuals [8]. Some other commensal strains of *E. coli* that do not cause diarrhoea are, however, known to produce a compound called 'colibactin' that is implicated in the causation of colorectal cancer. Not only in *E. coli*, but also in many other gut bacteria, we have to consider the possibility that

---

[1]The term 'Gram-negative' originally referred simply to the failure of cells to stain purple with Crystal violet ('Gram's stain') and this characteristic is now known to reflect bacterial cell surface structure.

some microbial strains may be both pathogenic and harmless, or even beneficial, depending on the individual and on the circumstances. This argument is well made by Martin Blaser in his book 'Our Missing Microbes' with reference to another member of the Proteobacteria, the stomach bacterium *Helicobacter pylori*. *H. pylori* has special attributes, such as the production of ammonia from urea, that enable it to survive in the highly acidic conditions on the wall of the stomach. This species has been identified as a pathogen that is a causative agent of stomach ulcers and it also contributes to stomach cancer [9]. Fortunately, it can be largely eliminated by treatment with a combination of antibiotics. On the other hand, it is argued that the presence of *H. pylori* is important in regulating immune responses. As a result, those without *H. pylori* may be more susceptible to asthma, allergy and other immune disorders such as coeliac disease than those who carry the bacterium [10].

By no means all gut Proteobacteria are as well studied as *E. coli*, but this phylum includes several other species that are known to play key roles in the gut microbial community. *Desulfovibrio* species, for example, have the ability to reduce sulfate to sulfide, which apart from being evil-smelling in its gaseous form, hydrogen sulfide (H_2S), can have toxic effects on the cells that line the gut wall. Another species of Proteobacteria *Bilophila wadsworthii* reduces sulfite to H_2S. The relative abundance of this species in the microbiome is increased on diets high in animal protein and fats (discussed in Chaps. 6 and 9).

Proteobacteria normally comprise a fairly small percentage (1–5%) of the microbiota of the healthy adult human gut, but this can increase considerably in disease states. Indeed, there is evidence that infection by pathogenic strains can actually change the gut environment so as to favour the pathogen [11].

Bacteroidetes

Bacteroidetes is one of the two most numerous phyla of bacteria colonizing the healthy human large intestine and typically account for 10–40% of total bacterial sequences detectable in faecal samples. In common with the Proteobacteria, they possess a Gram-negative cell surface with LPS, although their LPS is somewhat less immunogenic than that of pathogenic Proteobacteria. They tend to possess quite large genome sizes (around 6 million base pairs) by comparison with other gut bacteria and in several species these code for large numbers of genes concerned with the processing of carbohydrates of dietary and host origin (Chap. 7).

The two most common genera of gut Bacteroidetes are *Bacteroides* and *Prevotella*. In humans, there is increasing evidence that individual adults who possess high numbers of *Bacteroides* in their gut do not possess high numbers of *Prevotella*, and vice versa. The reasons for this intriguing variation between people are not yet fully understood, but we will consider some possibilities in Chap. 6. It is not yet clear whether on balance these bacteria are beneficial or detrimental for health and indeed this may depend on the species more than the genus profile. While most *Bacteroides* are considered harmless commensals, some species, notably

Bacteroides fragilis, include pathogenic strains. Several *Bacteroides* species have been reported as *opportunistic pathogens*, meaning that they may contribute to infections initiated by other organisms or under unusual circumstances where the host's defences against disease are compromised. As major fermenters of proteins, peptides and amino acids, *Bacteroides* can also produce toxic breakdown products that may promote colorectal cancer. Once again, the distinction between 'pathogen' and 'harmless commensal' becomes somewhat blurred. Another potentially harmful Gram-negative anaerobe *Fusobacterium nucleatum* (belonging to a distinct phylum, Fusobacteria) has been implicated in causation of colorectal cancer.

Bacteroidetes are obligate anaerobes, although they display a range of sensitivities to oxygen. While *Prevotella* spp. from the rumen are highly oxygen-sensitive, human colonic *Bacteroides* spp. can survive exposure to air for several hours [12].

Firmicutes

Firmicutes is a diverse phylum that can account for anything from 20 to 70% of bacteria in the healthy adult human gut, meaning that it and the Bacteroidetes are generally the two dominant phyla. Most (although not quite all) Firmicutes possess Gram-positive cell surface structures that consist of a single proteolipid membrane and an outer cell wall rich in a compound called lipo-teichoic acid. As it happens, not all bacteria with this surface structure actually stain Gram-positive (but this detail only matters to the relatively small number of microbiologists who use classical microscopy!). Most (although not all) Firmicutes have genomes whose DNA is characterised by rather more A:T base pairs than G:C base pairs. This is often described as the DNA %G + C content, which is typically between 25 and 40% ('low %G+C') in most Firmicutes.

The great majority of strains and species of Firmicutes that inhabit our intestine are not considered harmful, and indeed many appear to be positively beneficial to health. The most numerous Firmicutes bacteria in the human large intestine are obligate anaerobes that include several of the most abundant species found in faecal samples from the majority of human adults so far studied. There are three particularly abundant groups of Firmicutes that produce butyric acid as their major fermentation product. As we will discuss later in more detail, butyrate is used as a source of energy by the cells that line the large intestine and has a role in suppressing inflammation, and in preventing colorectal cancer and colitis.

Faecalibacterium prausnitzii One major butyrate-producing species (probably soon to be multiple species) is *Faecalibacterium prausnitzii*, which is generally among the top three most abundant bacterial species detected in faecal samples from healthy human subjects. This means that many of us will have 10 billion of this one species per ml of our gut contents. On the other hand, *F. prausnitzii* numbers are reportedly very much lower in individuals who suffer from ileal Crohn's disease, a debilitating condition that is marked by severe inflammation of the lining of the gut.

There is evidence to suggest that *F. prausnitzii* normally acts to reduce inflammation in the large intestine, not only through the production of butyrate, but also via other mechanisms still to be clarified. This has led to considerable interest in the potential of *F. prausnitzii* as a *therapeutic* agent [13]. But the organism is still something of a mystery. *F. prausnitzii* was first described and classified as a species of *Fusobacterium*, but difficulties in isolating and growing this bacterium delayed research until quite recently, when the isolation of new strains and the application of 16S rRNA gene sequencing led to its reclassification. Although previously isolated on medium containing rumen fluid, *F. prausnitzii* grows well on the rumen fluid-free medium YCFA² which supplies the acetate that it requires for growth [14]. *F. prausnitzii* is an obligate anaerobe, but it possesses an unusual mechanism (extracellular electron transport) for taking advantage of low concentrations of oxygen in its environment [15]. This might possibly help it to colonize niches close to the intestinal wall, where oxygen is leaking across into the gut from the blood supply. Perhaps surprisingly for such an abundant species, *F. prausnitzii* strains show a rather limited ability to utilize diet-derived carbohydrates for growth, failing to grow on starch or hemicellulose [15]. *F. prausnitzii* is a somewhat deviant member of the Firmicutes as it has a relatively high DNA %G + C content and its cell membrane organisation appears closer to that of Gram-negative than Gram-positive bacteria.

Eubacterium rectale *and* Roseburia *Species* Another group of butyrate-producing Firmicutes bacteria includes the highly abundant species *Eubacterium rectale*, together with the closely related *Roseburia* species [16]. These bacteria play a major role in fermenting undigested dietary carbohydrates that reach the large intestine to produce butyrate. We know this because when human volunteers change from normal diets high carbohydrate and fibre to carefully controlled medium or low carbohydrate weight loss diets, we see a sharp decrease in the amount of butyrate produced as a proportion of total fermentation acids in faecal samples. This is paralleled by a sharp decrease in the populations of *Roseburia* and *E. rectale* bacteria [17]. These diets had a more limited impact on *F. prausnitzii* numbers. *Roseburia* and *E. rectale* strains are among the few Firmicutes known to carry *flagella* that allow them to be motile, a characteristic that they share with Gram-negative pathogens such as *E. coli*. We know that the immune system has the ability to recognise flagellar proteins as part of its defence against infectious bacteria, but it is not yet clear exactly what impact these commensal flagellin proteins have on the host's immune system.

Lactate-Utilizing Firmicutes Other species among the Firmicutes have the capacity to produce butyrate from carbohydrates, or in some cases from amino acids

---

²'YCFA' stands for yeast extract, casitone and fatty acids (notably acetate, which is required for optimal growth of *F prausnitzii*); these are the main ingredients of the medium along with mineral salts and vitamins and an appropriate carbohydrate energy source (as detailed in Lopez-Siles et al. [14]). This medium was also used with great success by Browne et al. [22] to isolate a wide range of 'unculturable' human intestinal bacteria.

[18, 19]. Of particular interest, however, are those such as *Anaerobutyricum* (formerly *Eubacterium*) *hallii* and *Anaerostipes* species that can produce butyrate from lactate and acetate [20] (Fig. 4.4). There is a suggestion that such organisms may play a vital role in stabilising the microbial community of the large intestine and buffering it against changes in acidity (pH). We know that farm animals (cattle and horses) can suffer from potentially fatal lactic acidosis induced by certain diets, for example diets that deliver high doses of readily fermented carbohydrate to the rumen or caecum (Chap. 4). What happens is that the production of acids from rapid fermentation starts to overcome the ability of the gut to regulate the pH of its contents, with the consequence that the gut becomes more acidic. This situation is exacerbated when the bacteria mainly responsible for fermentation are ones that produce lactate (lactic acid) since this is more acidic (technically, it has a lower *pKa*) than the other acidic products of fermentation. Accumulation of lactate lowers gut pH to the point where the normally dominant species within the microbiota cannot grow, and the main bacteria that can still grow are those 'lactic acid bacteria' that produce still more lactic acid. Acidosis therefore represents a downward spiral. In common with many organic compounds, lactic acid exists as two *stereoisomers* (mirror images of each other) termed L- and D-forms. The body itself produces L-lactic acid when oxygen is in short supply, as discussed earlier, but microbes produce both forms and any D-lactic acid present in the gut or in the bloodstream is therefore the product of micro-organisms. It is this microbially-produced D-lactate that is particularly toxic to mammals, acting as a neurotoxin.

Bacteria that consume lactate normally help to prevent lactate accumulation in the gut. Within a certain pH range (neutral or mildly acidic, around pH 6–8), almost of the lactate that is produced is quickly consumed by these bacteria and we do not see lactate as a major end product of the mixed microbial community. In the rumen, Firmicutes bacteria (notably *Megasphaera elsdenii* and *Coprococcus catus*) that produce propionate from lactate appear mainly responsible for 'mopping up' lactate and so helping to prevent acidosis. While these species also occur in the human large intestine, it is bacteria such as *A. hallii* that produce butyrate from lactate that are numerically more abundant and probably more important in humans. Nevertheless, as these various lactate-utilizing bacteria are also affected by decreasing pH, their ability to consume lactate decreases as the pH drops close to 5. In humans, acidic pH resulting from uncontrolled lactic acid formation is seen in the condition 'short bowel syndrome' that can follow some surgical resection procedures. This can be fatal, although it can be avoided by ensuring there is not too much fermentable fibre from the diet.

Ruminococcus *and* Blautia We are still discovering the remarkable diversity and variety of metabolic capabilities among Firmicutes that colonize the intestine. *Blautia* and *Ruminococcus* species are anaerobes that do not produce butyrate. Although previously classified as ruminococci, these two genera are now recognized to belong to two distinct bacterial families. *Blautia* species belong to the Lachnospiraceae, whereas true *Ruminococcus* species belong (together with *F. prausnitzii*) to the Ruminococcaceae. The genus *Blautia* includes some species known to be 'acetogenic'—meaning that they have the remarkable ability to gain energy by forming acetate from carbon dioxide and hydrogen (or formate) via the Wood Ljungdahl pathway. In the gut it appears that these acetogens (such as *Blautia hydrogenotrophica*) can form additional acetate through this route at the same time as using other energy sources. As a result of their activities less hydrogen and formate accumulates within the system, and there is an additional net gain of carbon derived from carbon dioxide (Fig. 4.3). Some *Blautia* species, such as *B. gnavus*, have the ability to utilize host-derived *mucin* as an energy source for growth.

Ruminococcus species are remarkable mainly for their highly complex and specialised enzyme systems that allow them to extract energy from dietary fibre sources. We explore these further in Chap. 7. The most common human gut species is *Ruminococcus bromii*, which is a specialist starch degrading organism with an exceptional ability to break down dietary *resistant starches*, even when present in the uncooked state [21]. Other human colonic *Ruminococcus* species specialise in utilization of β-glucans ('*R. bicirculans*') or cellulose (*R. champanellensis*).

As genome sequences have become available for anaerobic Firmicutes, it has been shown that many of them possess genes very similar to those previously shown to be responsible for the formation and germination of spores in many *Bacillus* and *Clostridium* species. This is leading to a complete rethink of the biology of many organisms previously considered to be 'non-sporing anaerobes'. Sporulation genes have now been identified in many Firmicutes bacteria from the human intestine [22] and sporulation was recently demonstrated directly in one such species, *Ruminococcus bromii* [23]. This has profound implications for the microbial ecology of these species. The potential for a 'strict anaerobe' to survive outside the gut and to transfer between hosts must be fairly limited because of the impact of oxygen exposure upon the viability of vegetative cells. Spores are resistant structures that withstand oxygen, temperature variation and other challenges, however, and so will greatly increase the potential for transmission and acquisition of strains that are capable of sporulating.

Lactic Acid Bacteria The Firmicutes phylum also includes many organisms that are able to grow in air or that are somewhat tolerant of oxygen. The latter include facultative anaerobes and 'micro-aerophiles', i.e. organisms that grow optimally at low oxygen concentrations. Among these, Lactobacilli, Streptococci and Enterococci are well represented among the small intestinal microbiota, but typically account for a smaller percentage of the large intestinal community. These genera are known as 'lactic acid bacteria', as most produce lactic acid and in many cases also acetate as their major product from anaerobic metabolism.

Lactobacillus species are supplied commercially as *probiotics* in a wide range of fermented foodstuffs, mostly 'live' yoghurts. The term probiotic implies a health benefit to the consumer and there have been many research studies conducted in support of claims for beneficial effects on immune function, the prevention of gut infections and of allergies [24]. But how might probiotics actually work? Critics point out that the strains supplied from the foodstuffs are often not from species that are normally resident in the gut and that the acidity of the stomach in any case inactivates most of the viable organisms that we consume. It is generally acknowledged that the likelihood of permanently changing the gut microbiota by introducing a 'foreign' strain is therefore very low. As against this, we know from work with pathogens that only a few cells have to survive passage through the stomach to be able to replicate further down the gut if conditions are favourable to them there. Furthermore, permanent colonisation is not a requirement for an introduced microorganism to have an impact on its host. An important mechanism of probiotic action is through 'sampling' of the gut microbiota by specialist immune cells that line, or protrude into, the small intestine. In this way a passing dose ('bolus') of probiotic bacteria could be recognized by the immune system and cause much wider changes through the whole body. Under normal circumstances *Lactobacillus* species are not associated with infection, although a few rare cases have been reported [25]. Most importantly, however, existing probiotic organisms have a long history of safe use in food products across the world and their sale in commercial products goes right back to the 1920s. Indeed, fermented foods that deliver live micro-organisms to the gut have formed part of the human diet for a great many centuries before that.

Clostridia The Firmicutes phylum also includes some serious pathogens, notably certain strains and species of *Streptococcus*, *Enterococcus* (which incidentally are also 'lactic acid bacteria') and *Clostridium*. *Clostridium difficile*, although a member of the gut microbiota, can cause a life-threatening infection that has been called 'antibiotic-associated diarrhoea' because of its tendency to occur in hospitalised patients treated with antibiotics. This is because the 'healthy' gut microbiota normally acts as a barrier in preventing *C. difficile* infection. Only when the density of this community is decreased and its balance disturbed by antibiotics does *C. difficile* have the chance to establish, and once having been disturbed the community does not recover easily to its original state before antibiotic treatment. In common with many *Clostridium* species, *C. difficile* is able to form resistant spores that make it even harder to eliminate the organism from an individual or population. A new and highly successful therapy, however, has proved to be the transfer of faecal matter from a healthy donor (so called faecal microbiota transfer) that can lead to the restoration of a healthy microbial community and the barrier to infection that it provides (Chap. 12).

By no means all *Clostridium* species are harmful to the host and many can be regarded as commensals that form part of the normal healthy gut microbiota. The classification of species assigned to genera such as *Clostridium* and *Eubacterium* is constantly under review, informed by newly available genome sequence information, with existing species constantly being assigned to new or different genera. Simplistic assumptions about what are 'good' or 'bad' bacteria are often based on 'guilt by association'. This make little very sense if such judgements are based only on historic taxonomic assignments. Certainly, not all lactic acid bacteria are automatically 'good' and not all Clostridia are 'bad'!

Actinobacteria

Actinobacteria is a phylum of bacteria that possess Gram-positive cell wall structure. They are sometimes called 'high %G+C Gram-positives' because they possess relatively high %G+C in their genomic DNA. The main genera represented in the human gut are *Bifidobacterium* and *Collinsella*. As noted in Chap. 4, bifidobacteria ferment sugars mainly to lactate and acetate via the unique 'bifid shunt' pathway. Actinobacteria also include filamentous soil bacteria such as *Streptomyces* that are important producers of antibiotics, although these are not prominent among the gut microbiota.

Certain Bifidobacterial strains are widely promoted as probiotics in live yoghurts and native gut bifidobacteria are also common targets of prebiotics (substances, mostly types of fibre, that are designed to favour the growth of beneficial organisms in the gut). In 1899 the French bacteriologist Henri Tissier working at the Pasteur Institute first isolated bacteria then referred to as *Lactobacillus bifidus*, and now known as *Bifidobacterium* spp., that showed highly unusual Y-shaped cells under the microscope. The benefits of a gut microbiota rich in such lactic acid bacteria were soon championed by the Russian microbiologist (and Head of the Pasteur Institute) Eli Metchnikov, largely based on the longevity of Bulgarians who consumed fermented milk products. In human adults, *B. adolescentis* and *B. longum* are often among the most abundant species detected among bacteria in faecal samples. It is in infants, however, that bifidobacteria are most dominant, and particularly so in the gut of breast-fed infants. Mother's milk contains human milk oligosaccharides (HMOs) which create a special niche for infant bifidobacterial strains that compete very well for these as growth substrates [26]. HMOs could be thought of therefore as natural *prebiotics*. Furthermore, as producers of lactate, bifidobacteria tend to acidify the pH of the infant gut, thus tending to curtail competition from other bacteria. Much less is known about the abundant member of the Actinobacteria, *Collinsella aerofaciens* although early reports suggest a possible link with rheumatoid arthritis based on the pro-inflammatory properties of the bacterial cell wall [27].

Verrucomicrobia

Around 10 years ago, an unusual bacterium, named *Akkermansia muciniphila* was isolated by selecting for the ability to break down mucin. This organism turns out to be the only representative of the minor phylum Verrucomicrobia in the human gut [28]. In recent studies increased relative populations of *A muciniphila* in stool samples have been correlated with improved metabolic health accompanying weight loss [29] although the species is also significantly increased in colorectal cancer patients [30]. Notably, *A. muciniphila* is also correlated with slow gut transit as measured by the Bristol stool score [31] which may help to explain these associations. As a specialist mucin degrader, *A. muciniphila* is likely to make up a greater proportion of the microbiome under conditions where fibre concentrations are relatively low, either because of highly efficient fibre utilization (as occurs with slow transit) or limited fibre intakes.

Archaea

Archaea are mainly represented in the human intestine by methanogens of the species *Methanobrevibacter smithii* that produces methane gas from hydrogen and carbon dioxide or from formate [32]. The less abundant species *Methanosphaera stadtmanae* can make methane from methanol. Only around 50% of human adults carry high populations of methanogens and produce significant amounts of methane, while non-methanogenic individuals show very low populations of methanogenic archaea [33]. In the absence of methane formation, hydrogen and formate resulting from fermentation can still be consumed by acetogenic bacteria to produce acetate. Because of the poor energy yield from methane formation, methanogens are only capable of slow growth rates and they may tend to persist in individuals whose rate of gut transit is relatively slow. While some correlations have been suggested between archaea and gut disease, these may reflect the relationship with gut transit rather than indicating direct causation [32]. Intriguingly, there is also some evidence that methane may itself act to affect peristalsis and promote slow gut transit, perhaps creating conditions that favour methanogen survival [34].

Eukaryotes

The human intestine can potentially harbour many different eukaryotic micro-organisms. Protists include some troublesome pathogens such as *Entamoeba* and *Cryptosporidium* that can cause diarrhoea when they establish in the gut. Gut fungi also include pathogens such as *Candida albicans*, although a wide variety of other fungi are probably harmless commensals or transients in the gut. Infections caused by microbial eukaryotes do not generally respond to the same antibiotics that are effective against bacteria and it can be more difficult to find agents that selectively inhibit the eukaryotic pathogen without adversely affecting the host. Fortunately, some specialist anti-protozoal and anti-fungal agents are available.

Conclusions

Our detailed knowledge of gut micro-organisms and their individual effects on the host still depends critically on having cultured representatives. Cultured isolates provide information on features relevant to their microbial ecology, including sensitivity to oxygen and pH, substrates that can be used as energy sources (Chap. 7) and metabolites that are produced (Chap. 9). Crucially, interactions with the host and its immune system can be defined and the potential for toxin production, tissue invasion and pathogenicity determined. In the case of potentially beneficial commensal bacteria, the scope for therapeutic or probiotic use of single strains and strain mixtures is being actively explored and tested. On the other hand, we also need to understand how the whole gut microbiome varies between individuals or how it behaves in response to dietary change or heath status. For this, culture-independent profiling of the community using molecular methods has proved revolutionary in advancing our knowledge and understanding. We will explore some of these new insights in the next chapter.

References

1. Rickard AV et al (2003) Bacterial coaggregation: an integral process in the development of multispecies biofilms. Trends Microbiol 11:94–100
2. Dominy SS et al (2019) *Porphyromonas gingivalis* in Alzheimer's disease: evidence for disease causation and treatment with small molecule inhibitors. Sci Adv 5:eaau3333
3. Bik EM et al (2006) Molecular analysis of the bacterial microbiota of the human stomach. Proc Natl Acad Sci USA 103:732–737
4. Wang M et al (2005) Comparison of bacterial diversity along the human intestinal tract by direct cloning and sequencing of 16S rRNA genes. FEMS Microbiol Ecol 54:219–231
5. Bown RL et al (1974) Effects of lactulose and other laxatives on ileal and colonic pH measured by radiotelemetry device. Gut 1:999–1004
6. Walker AW et al (2011) Dominant and diet-responsive groups of bacteria within the human colonic microbiota. ISME J 5:220–230
7. Rajilic-Stojanovic M et al (2014) The first 1000 cultured species of the human gastrointestinal tract. FEMS Microbiol Rev 38:996–1047
8. Chassaing B, Darfeuille-Michaud A (2011) The commensal microbiota and enteropathogens in the pathogenesis of inflammatory bowel diseases. Gastroenterology 140:1720–1728
9. Wu Q et al (2013) Association between *Helicobacter pylori* infection and the risk of colorectal neoplasia: a systematic review and meta-analysis. Color Dis 15:E352–E364
10. Blaser MJ (2014) Missing microbes: how the overuse of antibiotics is fueling our modern plagues. Henry Holt & Company, New York
11. Stecher B et al (2007) *Salmonella enterica* serovar *typhimurium* exploits inflammation to compete with the intestinal microbiota. PLoS Biol 5:2177–2189
12. Flint HJ et al (2007) Interactions and competition within the microbial community of the human large intestine: key links between diet and health. Environ Microbiol 9:1101–1111
13. Sokol H et al (2008) *Faecalibacterium prausnitzii* is an anti-inflammatory commensal bacterium identified by gut microbiota analysis if Crohn's disease patients. Proc Natl Acad Sci USA 105:16731–16736
14. Lopez-Siles M et al (2012) Cultured representatives of two major groups of human colonic *Faecalibacterium prausnitzii* can utilize pectin, uronic acids and host derived substrate for growth. Appl Environ Microbiol 78:420–428
15. Khan MT et al (2012) The gut anaerobe *Faecalibacterium prausnitzii* uses an extracellular electron shuttle to grow at oxic-anoxic interphases. ISME J 6:1578–1585

16. Aminov RI et al (2006) Molecular detection, cultivation and improved FISH detection of a dominant group of human gut bacteria related to *Roseburia* and *Eubacterium rectale*. Appl Environ Microbiol 72:6371–6376

17. Duncan SH et al (2007) Reduced dietary intake of carbohydrates by obese subjects results in decreased concentrations of butyrate and butyrate-producing bacteria in feces. Appl Environ Microbiol 73:1073–1078

18. Louis P, Flint HJ (2009) Diversity, metabolism and microbial ecology of butyrate-producing bacteria from the human large intestine. FEMS Microbiol Lett 294:1–8

19. Louis P, Flint HJ (2017) Formation of propionate and butyrate by the human colonic microbiota. Environ Microbiol 19:29–41

20. Duncan SH et al (2004) Lactate-utilising bacteria from human feces that produce butyrate as a major fermentation product. Appl Environ Microbiol 70:5810–5817

21. Ze X et al (2012) *Ruminococcus bromii* is a keystone species for the degradation of resistant starch in the human colon. ISME J 6:1535–1543

22. Browne HP et al (2016) Culturing of 'unculturable' human microbiota reveals novel taxa and extensive sporulation. Nature 533:543–546

23. Mukhopadhya I et al (2018) Sporulation capability and amylosome conservation among diverse human colonic and rumen isolates of the keystone starch-degrader *Ruminococcus bromii*. Environ Microbiol 20:324–336

24. Hill C et al (2014) The International Scientific Association for Probiotics and Prebiotics consensus statement on the scope and appropriate use of the term probiotics. Nat Rev Gastroenterol Hepatol 11:506–514

25. Vankerckhoven V et al (2007) Infectivity of *Lactobacillus rhamnosus* and *Lactobacillus paracasei* in a rat model of experimental endocarditis. J Med Microbiol 56:1017–1024

26. Underwood MA et al (2013) A comparison of two probiotic strains of bifidobacteria in premature infants. J Pediatr 163:1585–1591

27. Zhang X et al (2001) Enzyme degradation and proinflammatory activity in arthritogenic and non-arthritogenic *Eubacterium aerofaciens* cell walls. Infect Immun 69:7277–7284

28. Derrien M et al (2008) The mucin degrader *Akkermansia muciniphila* is an abundant resident of the human intestinal tract. Appl Environ Microbiol 74:1646–1648

29. Dao MC et al (2016) *Akkermansia muciniphila* and improved metabolic health during a dietary intervention in obesity: relationship with gut microbiome richness and ecology. Gut 65:426–436

30. Weir TL et al (2013) Stool microbiome and metabolome differences between colorectal cancer patients and healthy adults. PLoS One 8:e70803

31. Vandeputte D et al (2016) Stool consistency is strongly associated with gut microbiota richness and composition, enterotypes and bacterial growth rates. Gut 65:57–62

32. Lurie-Weinberger MN, Gophna U (2015) Archaea in and on the human body: health implications and future directions. PLoS Pathog 11:e1004833

33. Florin THF et al (2000) Shared and unique environmental factors determine the ecology of methanogens in humans and rats. Am J Gastroenterol 95:2872–2879

34. Pimental M et al (2006) Methane, a gas produced by enteric bacteria, slows down intestinal transit and augments small intestinal contractile activity. Am J Physiol Gastrointest Liver Physiol 290:1089–1095

Chapter 6
Variability and Stability of the Human Gut Microbiome

During natural childbirth the gut of the newborn infant becomes colonized with microorganisms that come mainly from the mother, although they can also come from the environment. In non-natural (i.e. caesarean, as opposed to vaginal) births there is less opportunity for the normal, messy routes of inoculation via faecal material and body fluids and the early colonisers are more influenced by the skin surface and outside environment [1]. Which organisms become established during early life depends on the subsequent feeding regime, with breast feeding leading to different gut microbiota profiles compared with formula-based bottle feeding. In breast-fed infants, the faecal microbiota generally becomes dominated by bifidobacteria that compete most effectively for the *human milk oligosaccharides (HMOs)* produced in the mother's milk [2]. In contrast, infants who are not breast-fed exhibit higher relative numbers of Firmicutes, Bacteroidetes and Proteobacteria and lower bifidobacteria [3]. The difference in gut *microbiota composition* between caesarian and naturally delivered babies is significant before 3 months of age, but begins to disappear thereafter [4, 5].

A major shift in the developing infant gut microbiota occurs with the introduction of solid food, at the time of weaning. At this point other bacterial species can establish that specialise in degrading and utilizing solid food residues, especially plant fibre. From then on, the microbial community changes slowly towards the adult state, with the Bacteroidetes and Firmicutes phyla of bacteria becoming dominant. Changes in the balance of the gut microbiota occur again in later life, with proportions of *Bacteroides* for example somewhat higher in the elderly. While microbiota profiles appear stable within elderly individuals over time, they vary considerably between individuals with diet, lifestyle, medication and frailty all likely to be factors [6].

© Springer Nature Switzerland AG 2020
H. J. Flint, *Why Gut Microbes Matter*, Fascinating Life Sciences,
https://doi.org/10.1007/978-3-030-43246-1_6

Individual Variability

Core Species Do most human adults share the same *core* species of bacteria within their microbiota? To what extent does gut microbiota composition depend on such factors as developmental history, human genetic variation, geography, ethnicity, lifestyle and diet? The arrival of culture-independent techniques for microbial community analysis has allowed these fundamental questions to be addressed with far more rigour than was possible previously. In the human populations that have been studied most intensively some 50–70 'OTUs' (Operational Taxonomic Units,[1] the approximate equivalent of species but defined by sequence data alone) are present in the faecal samples of at least half of all individuals [7, 8]. A high proportion of these dominant species has in fact been cultured [9]. Furthermore, the top ten or so species in abundance have been remarkably consistent between studies that have used different technical approaches, whether culture-independent or culture-based (see Table 6.1, which compares a metagenomic study [10] with a cultural isolation study [11]). So, this would suggest a cautious 'yes' to our first question—it does appear that there are 'core' species to be found in the majority of healthy adult humans. Many of these organisms were highlighted in our brief overview of the human gut microbiota in Chap. 5. Detailed investigations into 'core species' have however been limited mainly to individuals from the US, Europe and Japan. As we will see shortly, evidence particularly from rural communities in other parts of the world remains scarce and we may still be unaware of other dominant human colonic species that are important globally.

It is worth noting that, even though a dominant species is shared by many individuals, the particular strains of that species that predominate are likely to differ between individuals. We can say this because a mass of scientific research shows the enormous capacity of bacteria to adapt through the transfer of genes and the selection of mutational variants. Given the vast populations of dominant species within the large intestine, continual adaptation and genetic change towards strain variants that are better fitted to the precise gut environment within that individual is certain to occur. In this way, each individual ends up with their own unique set of strains within their gut microbiota.

Enterotypes Despite the existence of core species that are common to most individuals, there is considerable variation between individuals in the detailed make-up of their gut microbiota. Much of this is the result of variation in the proportions of different species and groups within the microbiota. We might assume that (as with many complex biological traits) such variation between individuals would be

---

[1]An 'OTU' is an operational taxonomic unit. DNA sequences (typically from 16S rRNA genes, amplified by PCR) are compared and organized into a phylogenetic tree based on their similarity to each other. Sequences with more than a certain degree of similarity (the 'cut-off') are grouped within the same OTU. The choice of cut-off determines the taxonomic level, but it is often chosen to correspond approximately to cultured species. A proportion of OTUs will be from organisms that have not been classified or cultured.

Table 6.1 Dominant bacterial species detected in human faecal samples by metagenomic sequencing and by anaerobic culturing: a comparison of two selected studies

Genus, species	Phylum	Metagenomic sequencing Zhernakova et al. [10] [1135 adults—Europe] % Sequences	Anaerobic culturing Moore and Moore [11] [88 adults—US, Japan, Africa] % Cultured isolates
1. *Bifidobacterium adolescentis*	Actinobacteria	9.5	3.3 (5)
2. *Eubacterium rectale*	Firmicutes	8.4	5.8 (3)
3. *Ruminococcus bromii*	Firmicutes	6.9	2.9 (7)
4. *Faecalibacterium prausnitzii*	Firmicutes	6.2	4.6 (4)
5. *Subdoligranulum variabile*	Firmicutes	5.8	a
6. *Ruminococcus* sp.	Firmicutes	5.7	1.1 (15)
7. *Bifidobacterium longum*	Actinobacteria	5.4	1.1 (16)
8. *Colinsella aerofaciens*	Actinobacteria	4.2	10.9 (1)
9. *Dorea longicatena*	Firmicutes	2.8	a
10. *Akkermansia muciniphila*	Verrucomicrobia	2.7	a

[a]These taxa were not yet described in 1997. Other 'top 10' cultured species were *Bacteroides vulgatus* (2), *Bacteroides fragilis* (6), *Peptostreptococcus productus* (8), *Bacteroides uniformis* (9) and *Bacteroides stercoris* (10). Allowance has been made for name changes (*Fusobacterium prausnitzii, Eubacterium aerofaciens,* in 1997)

continuous, around an average (mean) state. Perhaps surprisingly, this appears not to be the case. Rather, there is evidence that microbiota variation is somewhat discontinuous, suggesting that there may be different types of community. Such alternative states of the gut microbiota have been called *enterotypes*. It was initially proposed that three enterotypes can be distinguished (characterised by predominance of *Prevotella, Bacteroides* and *Ruminococcus* species, respectively), based on metagenomic analysis of faecal bacteria within human populations [12]. An analogy was even drawn by some between enterotypes and blood groups (the latter are wholly determined by an individual's genes and are therefore fixed for life). This proposal, which was made before the effect of diet upon gut microbiota was fully appreciated, has proved somewhat controversial and subsequent researchers have often not been able to replicate the three proposed enterotypes. On the other hand, there have been many reports that distinguish between individuals whose microbiota is dominated by *Bacteroides*, as opposed to those dominated by *Prevotella* (both of these genera belong to the Bacteroidetes phylum). In one US study this difference was suggested to correlate with diet, with protein and fat intakes being higher in the *Bacteroides*-predominant individuals and plant fibre intake higher in the *Prevotella*-predominant individuals [13]. In another study, a group of African children were *Prevotella*-predominant, whereas a group of Italian children were *Bacteroides*-predominant [14]. Again, there was a considerable difference in fibre intake between the two groups, with fibre intake higher in the African children. Clearly, if enterotypes are driven by, or influenced by, diet, then they cannot be features of an individual that are 'fixed' for life.

Interestingly, at least one study suggests there is a relationship between the rate of gut transit and the abundance of *Prevotella* in our faecal microbiota [15]. Gut transit can be estimated from the commonly used six-point scale for faecal consistency known as the Bristol Stool Score (BSS) (this can often be seen in medical offices as a series of graphic illustrations of stools). *Prevotella* are reported to show higher relative abundance at higher values of the BSS, corresponding to looser stools, whereas the second 'enterotype' (termed *Bacteroides-Ruminococcus* by the paper's authors) is more associated with slow transit, or low BSS (harder stools). Since fibre is generally associated with faster gut transit, this might neatly explain the proposed association between fibre intake and a *Prevotella*-predominant microbiota.

Despite there being a reasonable degree of consensus among researchers that *Prevotella*- and *Bacteroides*-dominated microbiota can be distinguished within human populations, even here the term 'enterotype' may be misleading. A comprehensive analysis failed to find other bacterial taxa that were consistently associated either with *Prevotella* or with *Bacteroides* in faecal microbial communities worldwide and could show no clear separation of community profiles [16]. Nevertheless, it appears that within human populations the two genera can be useful as indicators (biomarkers) of responses to dietary intervention. For example, the faecal *Prevotella* versus *Bacteroides* status appears to be a good predictor of weight loss in volunteers when following low energy diets high in fibre, with *Prevotella*-predominant individuals losing the most weight [17]. *Stratification* according to *Prevotella* abundance within the microbiota may also help to predict responses to prebiotics. Overall, it is not clear whether predominance of *Prevotella* within one's gut microbiota offers a net heath benefit, there being evidence for both beneficial and detrimental effects [18].

To summarise, variations in microbiota composition between adults within human populations appear to be somewhat 'lumpy' rather than being smooth and continuous. The reasons for this are still not entirely clear. The explanation may lie with microbial interactions (inhibition, competition and cooperation) that make some configurations of the community intrinsically more stable than others. The particular configuration may then be determined by how the system responds to external factors that include gut pH, transit rates and dietary intake. These considerations are well illustrated by observations on the less complex microbiome found in the vagina, which exists as alternative types of community [19]. While the term 'enterotype' has been used to describe alternative states of the human gut microbiota, however, it is far from clear that enterotypes can be defined that apply across all human populations. It seems that we need a much better understanding of what drives overall system behaviour in gut microbial communities (Chap. 11).

How Diet Alters Microbiota Composition

We may each have our own distinctive microbiota profile, but does this remain constant through adult life, or does it change, in particular according to what we eat or what medication we receive? The only way to discover how diet affects the gut microbiota is by doing carefully controlled studies with human volunteers. The easiest design is probably the supplementation study, in which people follow their normal diet and lifestyle whilst also consuming a regular fibre supplement provided by the investigators. It is also possible however to conduct more precise studies in which all dietary intake is completely controlled over a period of time. This has the advantage that overall dietary intake can be accurately monitored, and that variations between people that might otherwise result from differences in their habitual dietary intakes are minimised. One such study compared the faecal microbiota profiles of overweight male volunteers when they were consuming a diet high in resistant (i.e. largely non-digestible) starch (RS^2) or when they were consuming an otherwise identical diet high in wheat bran. A 'cross-over' design was used so that each individual consumed the same diets, but at different times [9]. This investigation gave rise to several important conclusions. Community profiling based on amplified 16S rRNA gene sequences identified diet responsive species of bacteria, with different species responding to the RS and to the wheat bran diets within the same individuals. Interestingly, most of the responders were Firmicutes bacteria, with the species *Ruminococcus bromii* and *Eubacterium rectale* showing the greatest responses to the RS diet [9, 20]. Responses occurred rapidly within a few days of the diet switch and were then reversed by a subsequent switch to another diet within a similar time period (Fig. 6.1). Nevertheless, because a major part of the microbial community was unaffected by these dietary changes, the overall microbiota profile was most strongly influenced by the variation between individuals.

Another important study of this type looked at volunteers who were consuming either a 'plant-based diet', or a very extreme 'animal-based diet' that contained fat and protein but almost no plant fibre [21]. Changes in the faecal microbiota were

---

[2]Dietary 'resistant' starch (RS) refers to that fraction of dietary starch that is poorly degraded, or undegraded, by our own digestive enzymes (salivary and pancreatic amylases) in the upper gut. It therefore arrives in the large intestine, where it is available for fermentation by the gut microbiota (discussed further in Chap. 7). While some starches are resistant in their native state because of their particle structures, starches can also be rendered more, or less, resistant by cooking, processing or chemical cross-linking. The study referred to here used a commercially available type 3 (retrograded) resistant starch.

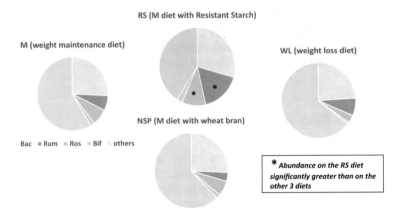

Fig. 6.1 Influence of diet on the human colonic microbiota. The average (mean) relative abundance of four bacterial groups (based on % 16S rRNA gene abundance) in faecal samples is shown for 14 overweight male volunteers who were following controlled diets. Data from Walker et al. [9]

again rapid, but more wide-ranging than was the case when only the fibre component was varied (as in the study described above). Bacteria increased in their relative abundance by the plant-based diet were mainly carbohydrate-utilizing Firmicutes such as *Roseburia* and *Ruminococcus* species. Conversely, the animal-based diet resulted in higher proportions of species involved in the metabolism, of mucin, fat and protein, such as *Akkermansia muciphila*, *Bilophila wadsworthii*, *Bacteroides* species and Proteobacteria.

These and many other studies have shown clearly that diet, in particular dietary fibre, can change the species composition of the human gut microbiota [22]. Furthermore, these changes happen quickly and in a fairly reproducible manner. By this we mean that the same diet-responsive species tend to be promoted by a given type of fibre in different individuals. What is also clear, however, is that individuals who start out with different species profiles in their resident gut microbiota can show different responses to a change in dietary intake. In the first controlled dietary study mentioned above the main species seen to respond on the RS diet was *Ruminococcus bromii* [9]. In two people who lacked this species at the start of the study, however, *R. bromii* was still not detected on the RS diet, although certain other species increased. Remarkably, only these two people failed to show complete fermentation of the RS, suggesting that *R. bromii* plays an indispensable (or *keystone*) role in the breakdown of this type of starch. This shows that the composition of the microbiota can affect the breakdown of fibre and thus the overall recovery of energy from fibre. We will revisit this issue in Chap. 8.

Causes of Microbiota Instability

It is worth stressing that the microbial community in the human intestine is never actually in a 'steady state' where its composition is constant with time. Because we eat periodically, the energy supply to the colonic microbiota arrives in 'pulses' (or 'boluses'). The arrival of fermentable fibre in the proximal colon promotes microbial growth and fermentation, leading to a localised drop in gut pH [23]. Once the pulse has passed, however, growth of the fibre-degrading species will slow down, fermentation and organic acid production will decrease and consequently the pH will rise. This means that the community must be in a constant state of cyclical change. Nevertheless, faecal samples (which represent a sort of 'historical record' of previous growth surges) taken at approximately the same time on different days will often show similar microbiota profiles. This is fortunate, since a great many research studies have been based on analysis of a single faecal sample per individual!

We have seen that diet has considerable potential to alter the composition and activities of our gut microbiota. Dietary fibres alter microbiota composition because they promote growth of bacterial species differentially within the community. Conversely, anything that acts to inhibit microbial growth differentially will also alter microbiota composition. In fact, a vast array of compounds arising from the diet or originating in the gut have the potential to inhibit microbial growth. These include phytochemicals, bile acids and microbial fermentation acids. In addition, the microbial community is exposed to other potentially disruptive influences that can include drugs (such as antibiotics that inhibit bacterial growth) and infections.

Impact of Antibiotics and Medicines Antibiotics can cause major disturbance of the gut microbiota in the short term, depending on the particular antibiotic, or combination of antibiotics, used. Recovery of the microbial community following cessation of antibiotic use may take a considerable time and may not be complete, potentially leading to the loss of sensitive species from the community [24, 25]. One of the consequences of such disturbance can be to reduce the protection provided by the normal microbial community against a pathogenic bacterium that is itself resistant to the antibiotics employed. This is what happens in 'antibiotic-associated diarrhoea' which is particularly associated with *Clostridium difficile* infection. Unfortunately, once the infection is established, conditions in the gut are altered such that *C. difficile* remains dominant and the normal microbiota does not re-establish itself [26]. This makes conventional treatment problematic. Fortunately, re-colonisation of the gut with a healthy microbiota (via 'faecal microbiota transfer') has proved extremely successful in combating *C. difficile* infection (discussed further in Chap. 12).

It is also likely that many medicinal drugs, although not targeted against microorganisms, have selective effects on the microbiome. For example, the drug commonly used to treat type 2 diabetes, metformin, alters the microbiome. This is of course a complicating factor when analysing the possible association between type 2 diabetes and the gut microbiome. As it turns out, type 2 diabetics not receiving metformin also show altered microbiota composition, with reduced proportions of butyrate-producing Firmicutes [27]. Similar problems apply with analysis of other conditions such as inflammatory bowel disease, where drug treatment is a clinical priority. Indeed, there is evidence that many commonly used medicinal drugs apart from antibiotics have effects on the gut microbiome [28] although most of these have yet to be fully investigated.

Gut Disorders Most of us are happy not to give a second thought to our digestive systems, provided that they do their job and do not bother us. We become only too aware of them, however, when something goes wrong. A whole variety of complaints, from the merely inconvenient to the serious and life-threatening, are covered by the terms 'digestive disorders' or 'gut disorders'. Our gut microbiota are involved to a degree in most of these, through their interactions with food components, but in some cases micro-organisms are clearly the primary cause of the disorder. Gut disorders are often associated with instability in the gut microbiota that is reflected in changes in the species composition of the microbial community. Such changes may be consequences, or causes, of the gut disorder.

The most obvious gut disorders are diarrheal diseases. These can be caused by a whole range of infectious micro-organisms, including bacteria, fungi and protozoa, and by viruses that target the gut. Many of the bacterial pathogens are well-known and notorious. Many of these are not normally resident in the gut microbiome, but are ingested from the environment through contaminated water, food or dirt. *Vibrio cholerae* is a highly infectious bacterium that can be lethal to humans. It produces a potent toxin that causes acute watery diarrhoea and dehydration and the resulting disease, cholera, is frequently fatal if untreated. The life cycle of *V. cholerae* involves alternative hosts that include tiny marine crustaceans. Other virulent bacterial gut pathogens mentioned already (Chap. 5) include strains of *Escherichia coli*, *Salmonella*, *Shigella* and *Campylobacter*. Gut infections can radically alter the resident gut community because of their effect on the host immune responses and on the gut environment. *Salmonella* infection, for example, results in changes in the gut microbial community that involve depletion of resident butyrate-producing bacteria, a switch to fermentative production of lactate by host cells and an increase in oxygen and nitrate in the gut lumen [29, 30]. These changes all help to promote further proliferation of the pathogen. Some pathogenic bacteria occur as part of the gut community but are normally held in check by the other members of the community. A disturbance in the microbial community, caused for example by antibiotic treatment, can remove this protective 'barrier effect' and allow the pathogen to infect, as discussed above in relation to *Clostridium difficile* infections.

On the other hand, there are a great many gut disorders that are not caused by a single infectious organism. Indeed, diarrhoea can be caused by factors that have nothing to do with infections. High concentrations of dissolved molecules (e.g. salt, sugars) in the gut contents can cause osmotic diarrhoea in which water is drawn across the gut membranes and into the gut. This can be caused by a sudden high intake, or by failure to absorb, the molecules such as salt and sugars responsible for this osmotic effect.

Another whole group of gut disorders is explained by food intolerances that originate from a genetic predisposition or from events in the early development of the immune system. A good example is coeliac disease, in which individuals suffer an immune reaction to the gluten protein that is present in some cereal (e.g. wheat)-based foods. Lactose intolerance is another common example of a genetic predisposition, although here fermentation of undigested lactose by the resident gut microbiota is responsible for producing the symptoms. Indeed, symptoms of bloating and excessive gas production (flatulence) can occur in anyone who consumes very large quantities of non-digestible, but fermentable fibre. It is worth mentioning that these symptoms have been observed with high doses of prebiotic fibres such as inulin [31].

A group of particularly important and serious gut disorders are those referred to as inflammatory bowel diseases (IBD). These involve abnormal immune responses by the body to its resident gut microbiota, resulting in severe inflammation and damage to the gut lining. The two main types of IBD are quite distinct. Crohn's disease, which can affect the small intestine, has a major heritable component with numerous genes concerned with immune function known to increase susceptibility. Genetic factors appear less important in the second type of IBD, ulcerative colitis, which mainly affects the large intestine. Alterations in the composition of the gut microbiota are seen in both diseases. In Crohn's disease, these include major decreases in many Gram-positive Firmicutes such as *Faecalibacterium prausnitzii* and increases in some unusual strains of *Escherichia coli*. These changes may be consequences of the inflammatory responses of the host, of changes in gut transit and the gut environment, or perhaps of drugs commonly used to treat the condition [32]. Most significantly, however, there is evidence that bacteria such as *F. prausnitzii* may help to suppress inflammation, while *E. coli* strains may actively promote the disease state (Chap. 5).

A very common group of gut disorders referred to as irritable bowel syndrome (IBS) do not show severe inflammation and are distinct from the inflammatory bowel diseases. IBS covers a whole range of symptoms that involve gut discomfort and sensitivity and is likely to represent multiple disorders with different causation, rather than a single condition [33]. Some cases may turn out to be due to other conditions, such as food intolerance or undiagnosed coeliac disease. Others, particularly those involving persistent diarrhoea (IBS-D), appear to follow on after viral or bacterial infections. Some cases, where symptoms (e.g. flatulence, bloating, gut sensitivity) suggest excessive fermentation, respond to decreases in fibre intake or to non-absorbable antibiotics, although there is no one agreed treatment regime [33]. Thus, while disturbances involving the gut microbiota and their activities undoubtedly play a role in IBS, this role has still to be clarified.

Colorectal cancer is the third most common cause of cancer-related deaths worldwide. This disease develops over a long period of time. Although there is a genetic component, most cases are attributed to poor diet and lifestyle and changes in gut microbiota profiles have been reported in colorectal cancer patients [34]. The gut microbiota are thought to contribute both to causation and prevention. The activities of a few pathogenic species such as *Escherichia coli* and *Fusobacterium nucleatum* may be directly involved in promoting cancerous changes. More generally, the balance of protective and carcinogenic (cancer-causing) microbial metabolites appears crucial in determining the gut environment [35] as we will see in Chap. 9.

Bacteriophage Viruses   Viruses known as bacteriophages are associated with most gut bacteria. Lytic bacteriophages infect cells and reproduce rapidly within their bacterial host which is then destroyed by phage lysozymes to release hundreds of new phage particles. Temperate bacteriophages are able to temporarily suppress the expression of their lytic genes and instead insert their genetic information as DNA into the bacterial chromosome, where it exists in a dormant state as a *prophage*. The prophage DNA is reproduced along with the bacterial cell but can become induced from this dormant state, for example in response to environmental or nutritional challenges. The induced phage then becomes lytic, quickly reproducing itself and destroying the bacterial host. The idea of using lytic bacteriophages as an alternative to antibiotics for combatting infective organisms (*bacteriophage therapy*) has been around for a long time and remains a realistic possibility [36].

   It is easy to see that bacteriophages must be a significant factor in the population dynamics of gut bacteria. Despite decades of very detailed research on the many bacteriophages that infect *Escherichia coli*, we know remarkably little about the bacteriophages that infect the dominant anaerobic bacteria of the human gut. Major changes in the intestinal virome are found in inflammatory bowel disease patients, however, notably the expansion of a group of bacteriophages (Caudovirales). This raises the strong possibility that bacteriophages contribute to the alterations in the bacterial community that occur in these conditions [37].

Diversity of the Gut Microbiome

The overall diversity[3] of the gut microbial community can be estimated from the analysis of faecal samples. Using metagenomics, it is possible to estimate a total 'gene count' (a measure of species 'richness') for the community. In a survey of 292 human adults (169 obese and 123 non-obese) the gene count displayed a discontinuous distribution around two peaks, one with higher diversity (HGC) and one of lower diversity (LGC) (Fig. 6.2) [38]. Microbiomes associated with the HGC peak tended to be higher in butyrate-producing Firmicutes species such as *F. prausnitzii* and the methanogenic archaeon *M. smithii* compared with LGC microbiomes. Obese and non-obese individuals were represented in both diversity groups. LGC individuals were somewhat more likely to be obese and to show symptoms associated with the early development of type 2 diabetes ('metabolic

---

[3]Diversity refers here to the variety of different species present within an individual's gut microbial community ('alpha diversity'). The simplest measure is 'richness' (the number of species, or the gene count, in a community) while other more complex measures (e.g. the Shannon index) also take account of species abundance.

Fig. 6.2 Bimodal distribution of gut microbiome 'richness' among human volunteers. Gene number (= richness) plotted for 292 volunteers. (From Le Chatelier et al. [38])

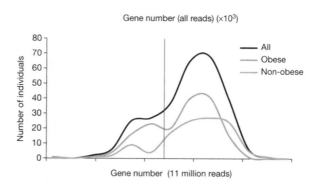

syndrome') but these differences were slight (e.g. mean BMI of 32 for LGC, 30 for HGC). Interestingly, in a separate dietary intervention study with obese volunteers, the gene count of LGC individuals increased towards the HGC value when these individuals were placed on a controlled diet [39]. One simple interpretation is that the LGC state is a consequence of habitually poor diets that are low in fibre. It is also possible that the composition of the LGC microbiota was directly contributing to poorer health status in these volunteers, although the dietary shift improved the health status of both LGC and HGC individuals.

Several studies have shown that 'hunter-gatherer' communities in Africa and South America possess significantly greater diversity in their faecal microbiota than urbanised populations in Western countries [40, 41]. Several factors are thought to contribute to this. For one thing, the diets of hunter-gatherers are probably a lot more varied than those followed by urbanised populations, including a wider variety of plant food sources that contain fibre and starches. Their food is not the product of industrial processing and food preparation and cooking methods will be determined by available equipment and fuel supplies. Furthermore, exposure to microorganisms in the environment is likely to be greater than among urban dwellers. For example, one study reported that food preparation by an African community used termite mounds, and some termite mound-associated microorganisms were detected among the faecal microbiota [14]. This all points to the likelihood that human communities with different diets and lifestyles and in different geographical locations may possess predominant species within their microbiota that differ from those in better studied, urbanised Western communities. As noted already, few of these bacteria have been studied in culture.

It tends to be assumed that this greater gut microbiota diversity of hunter-gatherers is 'natural' and is beneficial to health. The corollary of this assumption is that the relatively simplified 'Western' microbiota has lost critical members of our natural, or ancestral, microbial community and is less healthy [42]. An important element of this hypothesis is the alarming increase in antibiotic usage in many urbanised societies, particularly during very early stages of life [43]. This thesis is advanced most persuasively by Martin Blaser in his book 'Missing Microbes' [44, 45] and can be paraphrased as follows. Antibiotics have the potential to eliminate certain bacteria, including potentially beneficial species, from the gut and these may not easily be replaced later in life. Thus, as an adult, your microbiota may have been simplified as a result of exposure to antibiotics during early childhood, As a mother, you will then have less microbial variety in your gut to pass on to your own children through childbirth. If your delivery is via caesarean section, then even those micro-organisms that you do retain will have less chance to be passed on. So, the argument is that these factors may be leading to a progressive decrease in our gut microbial diversity through successive generations of modern living, with the loss of many species that are important to our well-being.

Further intriguing evidence on loss of microbiota diversity has come from studies with mice. It is possible to obtain *germ-free* rodents that carry no micro-organisms in their gut, and then to reintroduce particular strains, or whole communities into the pristine gut (they are then known as *gnotobiotic* animals). In one interesting study, gnotobiotic mice colonised with human faecal bacteria ('*humanised mice*') were fed diets either rich in, or depleted in, carbohydrates associated with plant fibre (referred to by the authors as 'high-MAC' or 'low-MAC' diets, respectively) [46]. The low-MAC diet resulted in a progressive loss of microbiota diversity, with *Bacteroides*-related bacteria being lost preferentially. Initially diversity loss could be reversed on switching back to the high-MAC diet, but with time diversity loss became irreversible and this low microbiota diversity was passed on from mothers to their offspring. It should be stressed that these experiments had to be performed in special facilities within 'isolators', meaning that the animals were not exposed to the outside environment.

It also seems reasonable to speculate that the variety of fibre types in the diet might influence microbiota diversity. An investigation using an in vitro chemostat model of the human gut microbiota concluded that the chemical complexity of individual fibre sources may be as important as the number of different fibres supplied [47]. In human volunteer studies we have found that matched diets with resistant starch as the main fibre resulted in lower microbiota diversity scores than those with wheat bran as the main fibre [20]. This may reflect the fact that starch is a simple homopolymer composed only of glucose, whereas wheat bran is chemically far more complex.

One likely consequence of reduced microbiota diversity is a loss of keystone species (mentioned earlier) that may play critical roles in the degradation of certain types of dietary fibre [48]. It seems likely that a diverse gut microbiota that retains such 'keystone' organisms will tend to maximise fermentation of fibre in the large intestine and the harvest of energy from the diet. This has been termed a 'permissive' gut microbiome. In contrast a less diverse gut microbiome typical of urbanised individuals (termed a 'restrictive' microbiome) lacks these organisms and so fails to ferment certain types of fibre. These different degrees of community diversity were proposed to explain the differing responses to dietary fibre supplementation observed between human dietary studies [49].

We should not leave this topic without thinking a bit more critically about the common assumption that gut microbiota diversity equates to health benefits. For example, the microbiota of breast-fed infants is almost certainly less diverse than that of bottle-fed infants, but we consider the former to be healthier. In other words, it is probably not the diversity index per se, but the balance of species within the microbiota that determines health. In adults it may be true that low gut microbiota diversity accompanies a poor diet that is low in plant fibre, and so is providing a valuable indicator of diet quality. Much more important, however, is the strong likelihood that certain bacterial species whose abundance is reduced in low diversity microbiomes (such as the butyrate-producing Firmicutes) are making a positive contribution to health.

Potential for Re-Colonisation

We have seen that gut disorders, antibiotic treatments and restricted diets all have the potential to alter the structure of the microbial community, including resulting in the loss of individual species and of overall diversity. A rather important question then is, can the 'missing' microorganisms be subsequently re-acquired and re-established? One presumed barrier to their acquisition is the oxygen sensitivity of many of the dominant gut bacterial species. This does not apply to facultative anaerobes, of course, which explains the ease with which we can pick up trouble-some infections caused by facultative pathogens such as *Escherichia coli* and *Salmonella* throughout life. But what about the dominant obligate anaerobes? Indeed, going back to our early development, how do we explain the apparent recruitment of so many new obligate anaerobes into the infant gut microbiota around the time of weaning?

For a long time most obligately anaerobic bacteria were classified as 'non-sporing', the exception being a few species of *Clostridium* (such as *C. difficile*) that are well known spore-formers. While there is no evidence for spore formation in most Gram–negative bacteria such as Bacteroidetes and Proteobacteria, it is now clear (as was noted in Chap. 5) that many anaerobic Firmicutes are likely to form spores. These organisms often possess an almost complete set of genes concerned with spore formation and spore germination within their genomes and in several species it has been demonstrated directly that spores are formed that can withstand exposure to heat and oxygen. In other species it may take time to discover the conditions required to promote spore formation and then to trigger germination, as these may not match those described in spore-formers studied previously, but the presence of the sporulation genes must be considered a very strong indicator [50]. What this means is that, in principle, our environment is likely to contain the spores of a wide variety of gut Firmicutes, thus increasing the chances that we will ingest these organisms. In addition, we know that some non-sporing anaerobes, including many *Bacteroides* species, are able to survive several hours of exposure to air. Survival outside the gut may also be enhanced by the carriage of bacteria in water droplets, and by their association with particles and with other bacteria. These factors will all tend to increase the potential for transfer of bacteria between individuals.

Initially, it was assumed that their oxygen sensitivity would preclude the use of commensal obligate anaerobes as probiotics or therapeutic agents. Now that we know spore formation is more widespread and with the development of new techniques for encapsulation [51] this looks much less of a problem. It is now clear that single strains or mixtures of strains ('cocktails') of anaerobes could be delivered to the gut successfully by oral inoculation, which opens up a host of possibilities and opportunities for redressing the balance of 'disturbed' colonic microbiomes that we will come back to later (Chap. 12).

Whether organisms can become established in the gut from an oral *inoculum* must depend on whether a vacant *niche* is available for the species to exploit. If the species is truly 'missing', then there must be a chance of there being a vacant niche. A neat example is provided by the bacterium *Oxalobacter formigenes*. This bacterium is a minor component of the healthy human gut microbiota that has the specialist ability to make use of oxalate in its metabolism and growth. When this species is missing in adults, these people have a greater risk of suffering from kidney stones, which are formed from calcium oxalate. Oral re-colonisation with *O. formigenes* was shown to successfully re-establish the bacterium and to decrease circulating oxalate levels [52]. Generally, the situation is far more complex than this, with multiple species competing for similar niches within the microbial community. A recent human study found that 30% of human volunteers could be colonised for at least 6 months by an introduced strain of *Bifidobacterium longum* (a species used in probiotics). Successful colonization coincided with there being low numbers of *B. longum* and, or, low numbers of genes concerned with utilization of specific carbohydrates (galactosides) initially in the individual's gut microbiota [53]. Again, this is consistent with the idea that an available niche is a requirement for establishment within the community.

Conclusions

Sequence-based analyses of the gut microbiota have enabled detailed descriptions to be made of gut microbial communities to an extent that was almost unimaginable 25 years ago. This has been profoundly important in helping to document the impact of diet, medication and individual variation on the gut microbiome. At the same time, it has generated hypotheses based on the linkage between microbiome profiles and many aspects of human health. At this point it becomes crucial to understand the real significance of alternative microbiome profiles. The term *dysbiosis* is often employed to refer to abnormal or altered states of the gut microbiome, implying departure from the 'normal' healthy balance. Such is the range of normality and the extent of inter-individual variation for microbiota composition, however, that the term defies precise definition. It follows that we need to move beyond simple sequence-based descriptions of microbial communities. We need to understand better how the different groups of gut micro-organism and their associated biochemical activities interact within the microbiome and what impact they have upon the host (ourselves). This will be the main topic of the chapters that follow.

References

1. Dominguez-Bello MG et al (2010) Delivery mode shapes the acquisition and structure of initial microbiota across multiple body habitats in newborns. Proc Natl Acad Sci USA 107:11971–11975
2. Zivkovic AM et al (2011) Human milk glycobiome and its impact on the infant gastrointestinal microbiota. Proc Natl Acad Sci USA 108:4653–4658
3. Harmsen HJM et al (2000) Analysis of intestinal flora development in breast-fed and formula-fed infants using molecular identification and detection methods. J Pediatr Gastroenterol Nutr 30:61–67
4. Rutayisire E et al (2016) The mode of delivery affects diversity and colonisation patterns of the gut microbiota during the first years of infants' life: a systematic review. BMC Gastroenterol 16:86
5. Akagawa S et al (2019) Effect of delivery mode on gut microbiota in neonates. Ann Nutr Metab 74:132–139
6. Claesson MJ et al (2012) Gut microbiota composition correlates with diet and health in the elderly. Nature 488:178–184
7. Tap J et al (2009) Towards the human intestinal microbiota phylogenetic core. Environ Microbiol 11:2574–2584
8. Qin J et al (2010) A human gene catalogue established by metagenome sequencing. Nature 464:59–65
9. Walker AW et al (2011) Dominant and diet-responsive groups of bacteria within the human colonic microbiota. ISME J 5:220–230
10. Zhernakova A et al (2016) Population-based metagenomics analysis reveals marker for gut microbiota composition and diversity. Science 352:565–569
11. Moore WEC, Moore LH (1995) Intestinal floras of populations that have a high risk of colon cancer. Appl Environ Microbiol 61:3202–3207
12. Arumugam M, Raes J, Pelletier E, Le Paslier D, Yamada T et al (2011) Enterotypes of the human gut microbiome. Nature 473:174–180
13. Wu GD et al (2011) Linking long-term dietary patterns with gut microbial enterotypes. Science 334:105–108
14. De Filippo C et al (2010) Impact of diet in shaping gut microbiota revealed by a comparative study in children from Europe and rural Africa. Proc Natl Acad Sci USA 107:14691–14696
15. Vandeputte D et al (2016) Stool consistency is strongly associated with gut microbiota richness and composition, enterotypes and bacterial growth rates. Gut 65:57–62
16. Gorvitovskaia A et al (2016) Interpreting *Prevotella* and *Bacteroides* as biomarkers of diet and lifestyle. Microbiome 4:15
17. Hjorth M et al (2019) *Prevotella*-to-*Bacteroides* ratio predicts body weight and fat loss success in 24-week diets varying in macronutrient composition and dietary fiber: results from a post-hoc analysis. Int J Obes 43:149–157
18. Precup G, Vodnar DC (2019) Gut *Prevotella* as a possible biomarker of diet and its possible eubiotic versus dysbiotic roles—a comprehensive literature review. Br J Nutr 122:131–140
19. Ravel J et al (2011) Vaginal microbiome in reproductive age women. Proc Natl Acad Sci USA 108(Suppl 1):4680–4687
20. Salonen A et al (2014) Impact of diet and individual variation on intestinal microbiota composition and fermentation products in obese men. ISME J 8:2218–2230
21. David LA et al (2014) Diet rapidly and reproducibly alters the human gut microbiome. Nature 505:559–563
22. Flint HJ et al (2014) The role of the gut microbiota in nutrition and health. Nat Rev Gastroenterol Hepatol 9:577–587
23. Bown RL et al (1974) Effects of lactulose and other laxatives on ileal and colonic pH measured by radiotelemetry device. Gut 1:999–1004

24. Dethlefsen L et al (2011) Incomplete recovery and individualized responses of the human distal gut microbiome to repeated antibiotic perturbation. Proc Natl Acad Sci USA 108(Suppl 1):4554–4561
25. Isaac S et al (2017) Short- and long-term effects of oral vancomycin on the human intestinal microbiota. J Antimicrob Chemother 72:128–136
26. Lawley TD et al (2012) Targeted restoration of the intestinal microbiota with a simple, defined bacteriotherapy resolves relapsing *Clostridium difficile* disease in mice. PLoS Pathog 8: e1002995
27. Forslund K et al (2015) Disentangling type 2 diabetes and metformin signatures in the human gut microbiota. Nature 528:262–266
28. Maier L et al (2018) Extensive impact of non-antibiotic drugs on human gut bacteria. Nature 555:623–628
29. Stecher B et al (2007) *Salmonella enterica* serovar *typhimurium* exploits inflammation to compete with the intestinal microbiota. PLoS Biol 5:2177–2189
30. Gillis CC et al (2018) Dysbiosis-associated changes in host metabolism generate lactate to support *Salmonella* growth. Cell Host Microbe 23:54–64
31. Ramnani P et al (2010) Prebiotic effect of fruit and vegetable shots containing Jerusalem artichoke inulin: a human intervention study. Br J Nutr 104:233–240
32. Hansen R et al (2012) Microbiota of de-novo pediatric IBD: increased *Faecalibacterium prausnitzii* and reduced bacterial diversity in Crohn's but not in ulcerative colitis. Am J Gastroenterol 107:1913–1922
33. Simren M et al (2013) Intestinal microbiota in functional bowel disease: a ROME foundation working team report. Gut 61:159–176
34. Wang T et al (2012) Structural segregation of gut microbiota between colorectal cancer patients and healthy volunteers. ISME J 6:320–329
35. Louis P et al (2014) The gut microbiota, bacterial metabolites and colorectal cancer. Nat Rev Microbiol 12:661–672
36. Schooley RT et al (2017) Development and use of personalised bacteriophage-based therapeutic cocktails to treat a patient with a disseminated resistant *Acinetobacter baumannii* infection. Antimicrob Agents Chemother 61:e00954–e00917
37. Norman JM et al (2015) Disease-specific alterations in the enteric virome in inflammatory bowel disease. Cell 160:447–460
38. Le Chatelier E et al (2013) Richness of human gut microbiome correlates with metabolic markers. Nature 500:541–546
39. Cotillard A et al (2013) Dietary intervention impact on gut microbial gene richness. Nature 500:585–588
40. Yatsunenko T et al (2012) Human gut microbiome viewed across age and geography. Nature 486:222–227
41. Schnorr SL et al (2014) Gut microbiome of the Hadza hunter-gatherers. Nat Commun 5:3654
42. Sonnenberg ED, Sonneburg JL (2019) The ancestral and industrialized gut microbiota and implications for human health. Nat Rev Microbiol 19:383–390
43. Bokulich NA et al (2016) Antibiotics, birth mode, and diet shape microbiome maturation during early life. Sci Transl Med 8:343ra82
44. Blaser MJ (2015) Missing microbes: how the overuse of antibiotics is fueling our modern plagues. Henry Holt & Company, New York
45. Blaser MJ (2017) The theory of disappearing microbiota and the epidemics of chronic diseases. Nat Rev Immunol 17:461–463
46. Sonnenberg ED et al (2016) Diet-induced extinctions in the gut microbiota compound over generations. Nature 529:212–215
47. Chung WSF et al (2019) Impact of carbohydrate complexity on the diversity of the human colonic microbiota. FEMS Microbiol Ecol 95:fly201
48. Ze X et al (2013) Some are more equal than others: the role of "keystone" species in the degradation of recalcitrant substrates. Gut Microbes 4:236–240

49. Wu GD et al (2016) Comparative metabolomics in vegans and omnivores reveal constraints on diet-dependent gut microbiota metabolite production. Gut 65:63–72
50. Browne HP et al (2016) Culturing of 'unculturable' human microbiota reveals novel taxa and extensive sporulation. Nature 533:543–546
51. Yeung TW et al (2016) Microencapsulation in alginate and chitosan microbeads to enhance viability of *Bifidobacterium longum* for oral delivery. Front Microbiol 7:494
52. Stewart CS et al (2004) *Oxalobacter formigenes* and its role in oxalate metabolism in the human gut. FEMS Microbiol Lett 230:1–7
53. Maldonado-Gomez MX et al (2016) Stable engraftment of *Bifidobacterium longum* AH1206 in the human gut depends on individualised features of the resident microbiome. Cell Host Microbe 20:515–526

Chapter 7
How Gut Micro-organisms Make Use of Available Carbohydrates

A significant fraction of the carbohydrates in our diet cannot be broken down by our own digestive enzymes. For simplicity, these 'non-digestible carbohydrates' are here referred to simply as 'fibre'. This diet-derived fibre passes through the stomach and small intestine to reach the large intestine, where it becomes the major energy source for the resident microbiota. Most of the fibre that we eat is fermentable, meaning that it can be broken down by our gut microbiota under the anaerobic conditions that prevail in the large intestine. On the other hand, some fibre is non-fermentable and behaves simply as a 'bulking agent', passing right through the gut un-degraded. In addition, carbohydrates produced by our own cells can be accessed by gut micro-organisms as energy sources.

Carbohydrates in Food

Most of the fibre in the human diet is non-digestible carbohydrate. Carbohydrates are so-called because their elemental composition consists of carbon and water (CH_2O) multiplied many times over (with loss of the odd water, as we will see). They fulfil many different roles in living organisms—but especially as building materials (cellulose in plant cell walls, wood), as energy stores (starch, glycogen, inulin), as readily transportable energy sources (sucrose, glucose, fructose) and as the backbone of nucleic acids (ribose in RNA, deoxyribose in DNA).

The common building block of carbohydrates is the single sugar molecule, which in nature typically contains five or six carbons. These sugars nearly always exist as ring structures. Glucose ($C_6H_{12}O_6$) is a six-carbon sugar in which five of the carbons and one oxygen form the ring while the sixth carbon forms a branch (Fig. 7.1). Glucose molecules can be linked together through the loss of water from adjacent hydroxyl (OH) groups on different molecules—this leaves them joined together via a single oxygen, forming a *glycosidic linkage*. This process of condensation can result in chains from two to many hundreds of glucose molecules in length. Long chains

© Springer Nature Switzerland AG 2020
H. J. Flint, *Why Gut Microbes Matter*, Fascinating Life Sciences,
https://doi.org/10.1007/978-3-030-43246-1_7

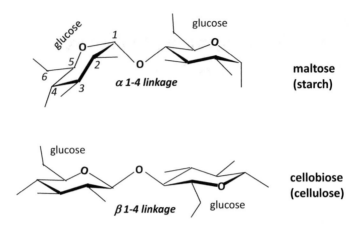

Fig. 7.1 Spot the difference. In **starch**, the repeating unit (maltose) consists of glucose sugars joined by alpha (α) glycosidic linkages. In **cellulose**, the repeating unit (cellobiose) consists of glucose sugars joined by beta (β) glycosidic linkages. This apparently minor difference explains the very different shapes and behaviour of starch and cellulose chains (see text). (nb. *O* oxygen. Following a common convention, the individual carbon, hydrogen atoms and OH groups in glucose are not detailed here. The numbering of the six carbons is however shown for the first glucose)

are referred to as *polysaccharides* (whereas glucose is the *monosaccharide*). What might appear trivial differences in polysaccharide structure can have consequences of huge biological significance. Thus, joining the adjacent glucoses between carbons 1 and 4 in one way produces an alpha-(1-4)-linkage, resulting in the polysaccharide starch (Fig. 7.1). We are very familiar with the fact that most starches can be digested thoroughly in our small intestine and that starch (largely from cereals and root vegetables such as potatoes) is a major source of energy in the human diet. Starch breakdown by our digestive enzymes (amylases) involves *hydrolysis*, i.e. the re-addition of water to destroy the glycosidic linkages, resulting in the release of sugars, which are mostly either glucose by itself or maltose (made up of two glucoses).

Simply joining adjacent glucoses between carbons 1 and 4 in the opposite way, however, produces a beta-(1-4)-linkage, resulting in the polysaccharide cellulose. This polysaccharide cannot be broken down at all by our own digestive enzymes and is one of the most durable organic molecules found in nature. Cellulose fibrils play a crucial role in providing rigidity to plant cell walls and so to the structure and architecture of the plant itself. In many land plants cellulose is combined with the non-carbohydrate compound lignin to form wood. Woody structures enable the tallest tree species to reach 100 m in height. The durability of cellulose has been of immense value to man through timber, clothing (cotton) and paper among a myriad of applications. The enzymes (cellulases) that can hydrolyse these beta-(1-4)-linkages most effectively are those produced by micro-organisms.

Why should the presence of alpha-(1-4) versus beta-(1-4) linkages in glucose chains make such a difference? A large part of the answer lies in the effect that they have on the shape adopted by the 'poly-glucose' chain. In starch, the alpha-(1-4)

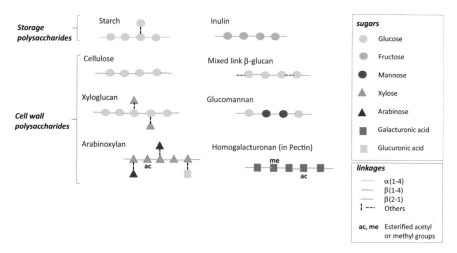

Fig. 7.2 Some plant-derived polysaccharides in food

chain tends to adopt a helical shape. Starch chains can align to form compact crystalline structures that do not dissolve in water at normal temperatures. Starch can nevertheless become 'gelatinous' when heated above 70 °C. The fact that the physical state of starches is greatly modified by temperature makes cooking a critical factor influencing the ability of starch-degrading enzymes to degrade starch in food. As a result, raw potato starch is almost completely non-digestible by our own enzymes, whereas cooked potato starch is 95% digestible. A variable fraction of starch molecules (depending on the plant species and variety) contain some alpha-(1-6) linkages that result in side branches; these also alter the overall solubility and digestibility of the starch.

In cellulose, the beta (1-4) linkages result in alternating orientations for the glucose molecules within the chain. The result is a straight chain with much greater rigidity. Furthermore, cellulose chains readily align with each-other through electrical attractions to form crystalline arrays and fibrils that are completely insoluble in water. The strength and durability of cellulose fibre is evident from the properties of cotton, paper and wood.

Whereas starch and cellulose are both composed simply of glucose, many other polysaccharides are chemically far more complex. In hemicelluloses, which like cellulose are important components of plant cell walls, chains of one sugar often provide a backbone that can link to other types of sugar (Fig. 7.2). Arabinoxylan, for example, consists of a chain of xylose sugars with arabinose sugars as side groups; some of the arabinose groups are also involved in ester linkages to lignin. Other polysaccharides that contribute to the plant cell wall are polymers of mannose, xylose, glucose in different combinations and with different linkages. By far the most complex group of plant structural polysaccharides are the pectins, with 21 different chemical linkages present in just one of its components, rhamnogalacturonan II [1].

Insoluble and Soluble Fibre

Much of the fibre that arrives in the human colon is insoluble in water. Insoluble fibre (plant cell fragments and starch particles) presents a particularly tough challenge for microbial attackers. The microbial community that is closely associated with fibre particles in gut contents differs in its species composition from that found in the liquid phase (Chap. 3). Specialist microbes that colonise insoluble fibre have first to become closely attached to the surface of the fibre particle and then be to erode or penetrate any protective surface layers. They have then to remove matrix material such as glucans and arabinoxylans. After that, accessing insoluble cellulose involves disrupting crystalline arrays to release soluble fragments that can enter the active sites of the relevant degradative enzymes. Storage polysaccharides such as starch and inulin are formed inside the plant cell, so that the surrounding plant cell wall polymers have to be disrupted before they can be accessed. Only once these initial stages of disruption and degradation have occurred can soluble carbohydrates be released that are then available for transport into the microbial cell. At this stage, many less specialised micro-organisms within the community can potentially enjoy a free lunch. Consequently, there is intense competition within the microbial community for the soluble breakdown products of insoluble fibres that results in the cross-feeding of partial breakdown products from primary degraders of insoluble fibre to other bacteria capable of utilizing these products. For example, while many human gut bacteria can grow when provided with soluble starches, only a few (notably *Ruminococcus bromii* and some *Bifidobacterium* species) seem able to attack insoluble starch particles [2]. Other species can however benefit from the initial degradation by *R. bromii*, in part because *R. bromii* does not take up some of the products, including glucose itself (Fig. 7.3). The ability to utilize different types of fibre is

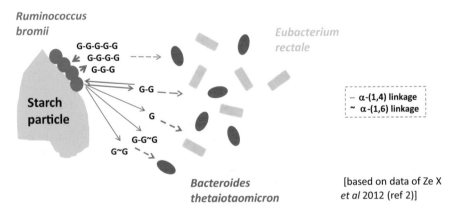

Fig. 7.3 Cross-feeding of break-down products from resistant starch. Breakdown products of starch range from glucose (G) and maltose (G–G) to oligosaccharides of 3 or more glucose molecules, and products with α(1-6) linkages. Not all are efficiently taken up by the primary degrader (*R. bromii*). Broken arrows indicate cross-feeding to other species

determined by the complement of degradative enzymes coded for by the genome. Also, as we will see, by the organization of these enzymes and the transport systems that import polysaccharide breakdown products into the cell.

Isolated strains can be assessed for their ability to use specific carbohydrates or plant structures. If the carbohydrate is soluble, then growth (increasing cell numbers) can be followed simply by measuring the turbidity of the culture in liquid medium. Insoluble substrates (which are themselves turbid) require other approaches, such as molecular detection of gene copy numbers (by qPCR) or the use of chemical assays to measure the disappearance of the substrate itself, or the increase in cell protein. Such information is relatively straightforward, but there are some important caveats. One is the choice of growth substrate. For example, is pig gastric mucin a suitable replacement for human colonic mucin (which is much harder to obtain)? How closely does a purified carbohydrate correspond to the fibre that the microbe actually encounters in the gut? For example, autoclaving is routinely used in microbiology to sterilise media, but autoclaving starch gelatinises it and makes it easier to break down than if it was raw or cooked by normal methods. Also, since we know that 95% of most starches are removed by our own digestive enzymes, it is really only relevant to examine the utilization of (non-autoclaved) resistant starches by human intestinal microbes [2]. Equally, while it is possible to obtain highly purified plant cell wall polysaccharides from chemical catalogues, it has to be kept in mind that in the gut they will be present, at least initially, interwoven with other polysaccharides and proteins within plant cell and tissue fragments. A second very important caveat is that we cannot be certain that all relevant micro-organisms within the in vivo microbiome will be capable of growth in the chosen laboratory medium.

'Carbohydrate-Active Enzymes' (CAzymes)

The variety of polysaccharide structures found in living organisms is matched by the diversity of enzymes that have evolved both to make them, and to break them down. These are referred to as 'Carbohydrate-active enzymes' or *CAzymes* for short. The vast numbers of CAzymes that are known (most of which are described only from DNA sequences) have been classified into 'families' that are based on similarities in their predicted protein sequences and structures (CAZy database, http://www.cazy.org/) [3]. Although these families do not tell us exactly what the enzyme does, a task that requires much careful experimentation, they can nevertheless provide strong clues.

Enzymes responsible for the manufacture (biosynthesis) of polysaccharides belong to one large group of CAZymes, the glycosyl transferases, which consist of 106 families (Table 7.1). Meanwhile, no fewer than 156 different families have been described for enzymes that break down glycosidic bonds by the addition of water (*glycoside hydrol*ases). Other groups of degradative enzymes are *polysaccharide lyases*, enzymes that employ an alternative mechanism for cleaving glycosidic binds that does not involve hydrolysis, and carbohydrate esterases, that remove groups that

Table 7.1 Carbohydrate-active enzymes described to date

Type of enzyme	Families	Role
Glycosyl transferases	106	Linking sugars together (biosynthesis)
Glycosyl hydrolases	156	Breaking glycoside links (by hydrolysis)
Polysaccharide lyases	29	Breaking glycoside links (not hydrolysis)
Carbohydrate esterases	16	Breaking ester linkages
Auxiliary enzymes	15	Carbohydrate degradation involving oxygenases

Information obtained from the Carbohydrate Active Enzyme database (http://www.cazy.org/)

are often found attached to carbohydrate chains by ester linkages. A further group of enzymes involved in oxidative cleavage of carbohydrate chains are mainly produced by fungi.

A high proportion of CAZymes within the families of glycoside hydrolase that are responsible for plant fibre breakdown are of microbial origin. The human genome itself codes for 98 glycoside hydrolases, including 7 that belong to the amylase family. However, humans lack enzymes that can degrade the polysaccharides that make up the bulk of plant fibre—specifically cellulose, beta-glucans, xylans, xyloglucans, mannans, galactans and pectins. These polysaccharides are major constituents of cereals, bran, vegetables, fruit, pulses and wholemeal bread. We also lack enzymes that allow us to degrade the storage polysaccharide inulin that is found in chicory and artichoke. As a result, all of this material passes un-degraded through the stomach and small intestine to arrive in the large intestine, where it provides a potential energy source for the resident microbial community. Even though we do produce our own starch-degrading enzymes, these still leave a fraction of dietary starch (resistant starch (RS)) un-digested. As a result, RS is present along with plant fibre in the digesta that enter the large intestine. Although it may represent as little as 5% of the total starch in the diet, average starch intakes are such that RS may often be one of the major energy sources for microbial growth in the colon [4].

Much research has gone into understanding how individual enzymes attack chemical linkages within single chains of sugar molecules. Equally important is to explain how these enzymes can gain access to dense, often crystalline, rafts or blocks of polysaccharide chains that lie in close alignment with each other, as in lignocellulose. Many of these enzymes are associated with protein sequences ('carbohydrate binding modules' or *CBMs*) that are essential for binding them to the substrate, in some cases also helping to open up the crystalline structure. In addition, different enzymes are known to act 'synergistically' to enhance each-others' action. For example, an enzyme that removes side branches often improves accessibility for another enzyme that attacks the main polysaccharide chain. Meanwhile, an enzyme that cuts within the main chain creates free chain ends that are more vulnerable to attack by 'endwise-acting' enzymes.

How Do Gut Bacteria Make Use of Plant Fibre?

Complex carbohydrates are generally the largest source of energy available to the microbiota of the large intestine and there is intense competition between different micro-organisms to gain access to them. Among bacteria, research has uncovered several distinct strategies for utilizing diet-derived carbohydrates, which offer particular advantages and disadvantages. The most abundant group of Gram-negative anaerobes in the human gut are the Bacteroidetes; human colonic *Bacteroides* species have the ability to utilize many different diet-derived carbohydrates for growth and many possess extraordinarily large numbers of genes coding for CAZymes in their genomes [5]. *B. thetaiotaomicron*, for example, has 261 GH genes while some other *Bacteroides* species possess even larger numbers (Fig. 7.4). The organization of glycoside hydrolase enzymes in *Bacteroides* species was elegantly revealed by the work of Abigail Salyers on the starch utilization system (Sus) of *B. thetaiotaomicron* (Fig. 7.5a). At first sight this organization is slightly puzzling because rather little amylase activity is produced outside the cell, and most of this activity resides in the space between the two membranes (the *periplasm*) or in the inner membrane. At the same time, several prominent 'Sus' proteins that are located in the outer membrane have no enzymatic activity, but these were nevertheless shown to be vital for starch utilization. It appears that the one outward-facing amylase clips soluble starch molecules into large fragments that are bound and transported though the outer membrane by these 'Sus' proteins [6]. The fragments are then further hydrolysed and assimilated once they have reached the periplasm. The key advantage of this system (which was termed a *sequestration mechanism* by Salyers) is that the bacterium efficiently isolates the products of hydrolysis for its own consumption, thus denying them to its immediate competitors. The key disadvantage is likely to be that the system is equipped to deal primarily with soluble carbohydrates, since there is currently no evidence that larger insoluble polysaccharides, let alone fibre particles, can pass through the cell's outer membrane. The various hydrolytic enzymes and Sus proteins, together with regulatory proteins that control when they are expressed, are coded for by clusters of genes in *Bacteroides* species. These are known as *PULs* (Polysaccharide Utilization Loci), with each PUL dedicated to a particular type of polysaccharide. The *B. thetaiotaomicron* genome contains 88 PULs of which 53 are dedicated to dietary carbohydrates and 35 to carbohydrates of host origin [7]. This gives the species an extremely wide repertoire of degradative enzyme systems and a sophisticated network of regulation that allows the bacterium to switch readily between the utilization of many different types of carbohydrate as they become available in the gut.

A completely different strategy is found in certain Firmicutes bacteria [8, 9]. Several species of *Ruminococcus*, including *R. flavefaciens* and *R. albus* from the rumen and *R. champanellensis* from the human colon, have the ability to degrade cellulose and to break down insoluble plant fibre. These species produce a wide array of hydrolytic enzymes, including those needed to degrade all the major constituents of the plant cell wall (cellulose, arabinoxylan, mannan, pectin). As these are Gram-

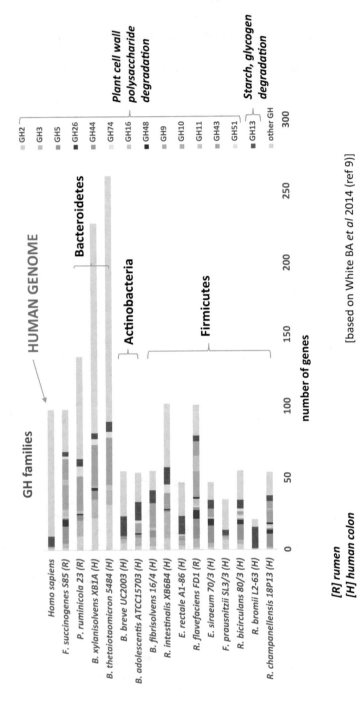

Fig. 7.4 Genomic capacity for dietary polysaccharide degradation—glycoside hydrolase (GH) genes

positive bacteria, there is no periplasm and almost all the degradative CAZymes are secreted to the outside of the cell. Importantly, however, these enzyme proteins mostly remain anchored to the surface of the cell. In *R. flavefaciens* and *R. champanellensis* many of them are assembled into large and elaborate enzyme complexes known as *cellulosomes* (Fig. 7.5b). Cellulosomes are organized through the attachment of enzymes to 'scaffolding' proteins that carry repeated regions known as *cohesins* and are attached to the bacterial cell surface. In the presence of

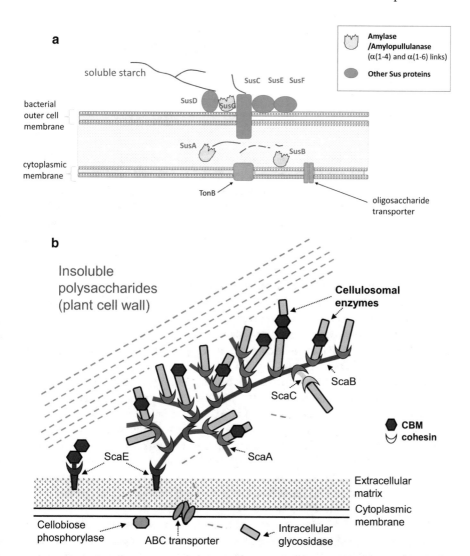

Fig. 7.5 (**a**) Starch-utilization system (Sus proteins) of *Bacteroides thetaiotaomicron* ("sequestration" system). (**b**) Cellulosome enzyme complex involved in plant cell wall breakdown by *Ruminococcus flavefaciens*. (**c**) Starch-degrading enzymes on the cell surface in *Ruminococcus bromii* ("amylosome" organization)

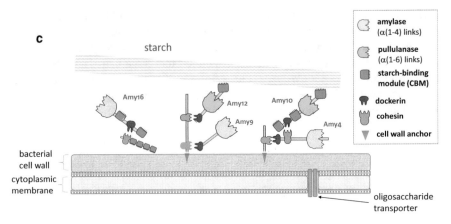

Fig. 7.5 (continued)

calcium, a *dockerin* sequence associated with an enzyme binds onto each of these cohesins. The *R. flavefaciens* genome codes for 220 presumed cellulosomal proteins that carry dockerins [10]. These include lyases, esterases, proteinases and many proteins unknown function, in addition to 59 glycoside hydrolases from 14 different GH families. This impressive battery of extracellular enzymes evidently gives the bacterium the power to cut through the complex structure of the cell wall and so to access sugars that it can use for growth. In the process of solubilizing the plant fibre, multiple breakdown products are released, some of which can be 'poached' by other species within the microbial community. Partly for this reason, perhaps, breakdown products tend to be transported into the cell as di- or oligo-saccharides rather than as single sugars in *Ruminococcus* species, many of which do not even transport glucose. These *Ruminococcus* species appear to be specialists in degrading insoluble plant fibre, but they require a wide range of enzymatic capabilities to attack the chemically complex plant cell wall.

The most abundant *Ruminococcus* species in the human colon, *R. bromii*, also specialises in degrading insoluble carbohydrates, but it targets particles of resistant starch rather than plant cell walls. This requires a far more limited enzymatic repertoire, since starch is composed only of glucose residues. Nevertheless, we find that five extracellular amylases in this species are organized into a cell surface complex, or *amylosome* (Fig. 7.5c) through the same mechanisms that are involved in the construction of the cellulosome [11]. Genes coding for glycoside hydrolases tend to be scattered around the chromosome in these *Ruminococcus* species, rather than being clustered into PULs as is found in Bacteroidetes.

Other solutions for polysaccharide utilization have of course evolved that lie between these two contrasting 'paradigms' of sequestration systems and extracellular enzyme complexes [8]. Another group of Firmicutes, those related to *Roseburia* and *Eubacterium rectale*, for example, clearly possesses PULs analogous to those in the Bacteroidetes [12]. While the main starch-degrading enzymes in *E. rectale* are expressed on the outer cell surface, there is also evidence for proteins that assist in capturing breakdown products [13]. The genomes of fibre-degrading Firmicutes generally possess lower numbers of CAZymes than those of the Bacteroidetes [4], but this may simply indicate that they are more specialised and have smaller genomes. There are strong indications that Firmicutes play important roles in the degradation of insoluble fibres such as wheat bran [14].

Host-Derived Carbohydrates

Polysaccharides fulfil some very important roles in mammalian cells and tissues, with glycogen (or 'animal starch') acting as the main energy store in muscle. Particularly important in the gut are the mucopolysaccharides, which together with protein components make up the mucin that protects the gut wall, especially in the large intestine. It is these polysaccharides (such as hyaluronic acid, a polymer composed of alternating glucuronic acid and *N*-acetyl glucosamine units) that make mucin slippery, so facilitating the passage of digesta down the gut. In addition, a great many mammalian proteins (termed glycoproteins) carry carbohydrate side chains.

As mentioned already, *Bacteroides thetaiotaomicron* has some 35 PULs in its genome that are devoted to the utilization of host-derived carbohydrates. Thus, it has the capacity to exploit as energy sources a very wide range of carbohydrate structures that are present on the surface of human cells, or in secreted mucin. There is evidence from mouse studies that, *B. thetaiotaomicron* switches over to using host-derived carbohydrates when dietary carbohydrates are in short supply by regulating expression of its PUL gene clusters [15]. This suggests that if we do not keep our gut bacteria supplied with fibre, they may end up feeding on us instead!

Certain sugars, such as fucose and sialic acid, are commonly found attached to human glycoproteins. Only some bacteria can use these for growth. Utilization of fucose requires a special pathway, the propanediol pathway, which has been found in some Gram-negative bacteria including *Salmonella* and in some Gram-positive Firmicutes species such as *Roseburia inulinivorans* [16].

Breast milk contains a special group of carbohydrates, human milk oligosaccharides (HMO), that do not get digested and so reach the infant large intestine intact [17]. A variety of HMOs are built up from lactose by the addition of other sugars, including fucose, sialic acid and N-acetyl glucosamine (Fig. 7.6). The same sugars and sugar chains are also found in human glycoproteins, so that HMOs act as mimics of glycans present on the surface of gut epithelial cells. This is significant because these structures are also used as targets by pathogenic bacteria and viruses, meaning that the HMOs act as 'decoys' that help to prevent the gut surface from binding these infectious agents. Glycoproteins such as lactoferrin produced in breast milk are also known to play an important role in preventing gut infections.

In addition, we know that HMOs are used as energy sources for growth by human gut bacteria. This growth stimulation by HMOs is very selective, however, with only

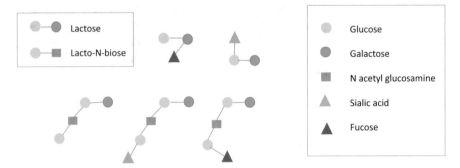

Fig. 7.6 Structures of some human milk oligosaccharides. A wide variety of compounds are produced from the disaccharides lactose and lacto-*N*-biose, with addition of fucose or sialic acid. Some of the smaller HMOs are shown here. (Based on Smilowitz et al. [17])

certain species of *Bifidobacterium* so far reported to specialise in their degradation and uptake. Interestingly, two different strategies appear to be employed. *B. longum* subsp. *infantis* transports the HMOs into the cell, then hydrolyzes them by means of the appropriate GH enzymes inside the cell. *B. bifidum* on the other hand exports enzymes that remove sialic acid and fucose, and then transports disaccharides into the cell for processing [17]. The growth promotion of specific bifidobacterial strains within the infant intestine by these oligosaccharides from maternal breast milk is therefore an entirely natural 'prebiotic' effect (discussed below). There is evidence for beneficial effects of these bifidobacteria in suppressing inflammation and improving barrier function (or preventing 'leakiness') of the gut lining. The amount and precise structures of HMOs produced in breast milk are influenced by the mother's genetic makeup, specifically their Secretor and Lewis blood types [17].

Dietary Fibre Intake and Prebiosis

Increasing the consumption of fibre is generally seen as beneficial to health in most human adults and this view is based on extensive analysis of a great many epidemiological studies [18, 19]. Fibre tends to speed up the passage of food through the gut, thus helping to prevent constipation. Increased intake of most types of fibre will increase microbial numbers and microbial activity in the large intestine. The increased formation of short chain fatty acids (SCFA) from carbohydrates via fermentation has multiple effects, one of which is to make the contents of the large intestine slightly acidic, at least in the proximal region where fermentation is most rapid. SCFA have beneficial effects on host physiology that we will consider in Chap. 9. Microbial breakdown of plant fibre also releases many non-carbohydrate phytochemical compounds that find their way into the bloodstream and some of these are considered beneficial to health.

Fibre intake also changes the species composition of the gut microbiota. Fibres therefore provide a potentially simple and 'natural' approach to manipulating our gut microbiota and its activities. But, the effects of fibre are not the same in everybody. A striking example is the fact that some two-thirds of all adults word-wide are intolerant of lactose and prefer to avoid dairy products. That is because lactose-intolerant people stop producing the human digestive enzyme that breaks down the sugar lactose in adult life, although they do produce this enzyme as infants. As a result, lactose then behaves as a fibre and arrives in the large intestine where it is fermented by the resident gut microbiota, resulting in problems of flatulence and digestive discomfort. On the other hand, in lactose-tolerant individuals (including many of Northern European descent) the ability to process lactose is retained into adulthood and dairy products are not a problem [20]. So, host genetics plays a crucial role in determining whether lactose behaves like a fibre for adult humans. Other groups of individuals can be intolerant of high dietary fibre intakes for a variety of different reasons. Thus, in some forms of irritable bowel syndrome (IBS) gastroen-terologists recommend a diet that is low in soluble fibre and non-digestible sugars (FODMAPS)[1] in order to minimise the symptoms of digestive discomfort. We should always bear in mind that nutritional recommendations based on the 'average individual' may not suit everyone.

At the risk of causing confusion, we should explain that, as a di-saccharide composed of two monomers (glucose and galactose), lactose does not fall within the current nutritional definition of a *fibre*, even though it behaves like one in many people. The current definition of fibre[2] includes "soluble and insoluble non-digestible carbohydrates with three or more monomers". As such, the definition also excludes another di-saccharide lactulose (derived from lactose) that is completely indigestible and is used medically as a laxative. Thus, not all non-digestible carbohydrates are formally classified as fibres.

Dietary manipulation involving specific fibres is referred to as *prebiosis* when the resulting changes are considered beneficial [21]. The original targets of prebiotic fibre tended to be bifidobacteria and the first prebiotic fibres were inulin or its derivative, fructo-oligosaccharides. There is good evidence that bifidobacteria can be promoted in this way, based on analysis of faecal samples taken following supplementation of adult diets with these prebiotics, but other questions remain to be answered. Specifically, what is the overall response of the community to the prebiotic, i.e. which other groups might be increased or decreased by its consump-tion? Now that we have the techniques to follow microbiota changes in detail, it has become clear that multiple, often unrelated, species can be promoted by a given

---

[1]'FODMAPS' stands for 'Fermentable oligosaccharides, disaccharides, monosaccharides and polyols'. This is in fact somewhat broader than the current definition of soluble fibre as it includes molecules such as lactose that have fewer than three sugar residues.

[2]Previously there was a tendency to distinguish between resistant starches and non-starch plant fibre, with the term fibre limited to 'non-starch polysaccharides (NSP)' such as cellulose, xylans, pectins and inulin. This has now been largely abandoned, with resistant starch being regarded as a fibre.

prebiotic [22]. If it is assumed that the prebiotic is associated with a health benefit, do we then know which species is responsible for the beneficial effects? For example, we now know that inulin can promote *Faecalibacterium prausnitzii* in addition to *Bifidobacterium* spp. in adult human volunteers [23], so which of these bacteria is more important for the prebiotic effect? We need to understand better how different prebiotics promote different groups and species within the microbiota. This must depend on the degradative capabilities of different species and strains within the microbiota, so that genome information combined with microbial culture, and dietary intervention and chemostat studies, are all providing new insights. In principle, it should become possible to select or design fibres with the aim of targeting specific groups of gut bacteria ('targeted prebiosis'). This could create some powerful new approaches for manipulating the species composition of the gut microbiota.

Another very important question is, how consistent are prebiotic effects across different individuals who receive the same dietary treatment? There is increasing evidence that an individual's microbiota response to dietary manipulation depends on the initial composition of their gut microbiota and on their own genotype. If the intended target organisms are absent, or fail to be promoted, we have to ask what other microorganisms may be promoted instead and whether are these likely to be beneficial or detrimental to health? Clearly there is much work to be done here.

Should prebiotics simply be regarded as simply a subset of dietary non-digestible carbohydrates (plant fibre and resistant starch) from which they are mostly derived? In some respects, the answer is yes—but there are some important caveats. For one thing, replacing complex natural dietary fibre with a single prebiotic might reduce microbiota diversity and lose other benefits of fibre intake such as the supply of phytochemicals. Also, when it comes to synthesising novel prebiotics, or modifying fibres (e.g. chemical cross-linking of resistant starches to make them more resistant to digestion) we are creating new fibres that require careful assessment of their impact in the gut.

Conclusions

Detailing the activities of isolated organisms and their enzyme systems has obvious value and fascination for researchers and helps to define the potential for carbohydrate utilization in vivo. This does not however allow us to predict the success of different microbial species within the complex microbiota to compete for, or respond to, fibre sources in the intestinal tract. The only definitive way to discover this is by doing human dietary studies, as we discussed in Chap. 6. Our intake of dietary fibre provides one of the most effective ways to manipulate the species composition and activities of our gut microbiota. Our understanding of which fibres promote different bacteria (and why) is increasing rapidly, as is our understanding of differences between individuals in their reactions to fibre intake. As we will see in the following chapters, however, many of the very real health benefits of fibre may derive from effects on the whole community and its metabolites, rather than on individual microbial species.

References

1. Ndeh D et al (2017) Complex pectin metabolism by gut bacteria reveals novel catalytic functions. Nature 544:45–65
2. Ze X et al (2012) *Ruminococcus bromii* is a keystone species for the degradation of resistant starch in the human colon. ISME J 6:1535–1543
3. Cantarel BL et al (2009) The Carbohydrate-Active EnZymes database (CAZy): an expert resource for glycogenomics. Nucleic Acids Res 37:D233–D238
4. Macfarlane GT, Englyst HN (1986) Starch utilization by the human large intestinal microflora. J Appl Bacteriol 60:195–201
5. El Kaoutari A, Armougom F, Gordon JI, Raoult D, Henrissat B (2013) The abundance and variety of carbohydrate-active enzymes in the human gut microbiota. Nat Rev Microbiol 11:497–504
6. Reeves AR, Wang GR, Salyers AA (1997) Characterization of four outer membrane proteins that play a role in utilization of starch by *Bacteroides thetaiotaomicron*. J Bacteriol 179:643–649
7. Martens EC et al (2009) Complex glycan catabolism by the human gut microbiota: the Bacteroides Sus-like paradigm. J Biol Chem 284:24673–24677
8. Flint HJ, Bayer EA, Rincon MT, Lamed R, White BA (2008) Polysaccharide utilization by gut bacteria: potential for new insights from genomic analysis. Nat Microbiol Rev 6:121–131
9. White BA et al (2014) Biomass utilization by gut microbiomes. Annu Rev Microbiol 68:279–296
10. Rincon MT et al (2010) Abundance and diversity of dockerin-containing proteins in the fibre-degrading rumen bacterium *Ruminococcus flavefaciens* FD1. PLoS One 5:e12476
11. Mukhopadhya I et al (2018) Sporulation capability and amylosome conservation among diverse human colonic and rumen isolates of the keystone starch-degrader *Ruminococcus bromii*. Environ Microbiol 20:324–336
12. Sheridan PO et al (2016) Polysaccharide utilization loci and nutritional specialization in a dominant group of butyrate-producing human colonic *Firmicutes*. Microb Genom 2:e000043
13. Cockburn DW et al (2015) Molecular details of a starch utilization pathway in the human gut symbiont *Eubacterium rectale*. Mol Microbiol 95:209–230
14. Duncan SH et al (2016) Wheat bran promotes enrichment within the human colonic microbiota of butyrate-producing bacteria that release ferulic acid. Environ Microbiol 18:2214–2225
15. Sonnenburg JL et al (2005) Glycan foraging in vivo by an intestine adapted bacterial symbiont. Science 107:1955–1959
16. Reichardt N et al (2014) Phylogenetic distribution of three pathways for propionate production within the human gut microbiota. ISME J 8:1323–1335
17. Smilowitz JT et al (2014) Breast milk oligosaccharides: structure function relationships in the neonate. Annu Rev Nutr 34:143–169
18. Reynolds AM et al (2019) Carbohydrate quality and human health: a series of systematic reviews and meta-analyses. Lancet 393:434–445
19. Aune D et al (2011) Dietary fibre, whole grains and risk of colorectal cancer: systematic review and dose response analysis of prospective studies. Br Med J 343:d6617

20. Udigos-Rodriguez S et al (2018) Lactose malabsorption and intolerance: a review. Food Funct 9:4156–4068
21. Gibson GR et al (2004) Dietary modulation of the human colonic microbiota: updating the concept of prebiotics. Nutr Res Rev 17:259–275
22. Chung WCF et al (2016) Modulation of the human gut microbiota by dietary fibres occurs at the species level. BMC Biol 14:3
23. Ramirez-Farias C et al (2009) Effect of inulin on the human gut microbiota: stimulation of *Bifidobacterium adolescentis* and *Faecalibacterium prausnitzii*. Br J Nutr 101:541–550

Chapter 8
Do My Microbes Make Me Fat? Potential for the Gut Microbiota to Influence Energy Balance, Obesity and Metabolic Health in Humans

In the last two chapters, we have seen that the composition of the gut microbial community responds to dietary change and that the microbiota contribute to the fermentation and transformation of dietary fibre. The gut microbiota also process other dietary components (fats and proteins) and interact with host cells, giving them the potential to influence many aspects of human nutrition and health. The possibility that our gut microbiota influence obesity and metabolic disease (diabetes) has been particularly topical in view of the increasing challenges posed by these conditions. We will start here by looking at the direct contribution of our microbiota to the recovery of energy from the diet, before moving on to consider the more complex interactions that occur with human physiology and metabolism.

Energy Harvest via Short Chain Fatty Acids

Short chain fatty acids (SCFA) produced by the intestinal microbiota have multiple effects on host physiology and metabolism that we will consider further in Chap. 9. For the moment, we will focus on the contribution made by SCFA to host cells as energy sources. In humans, butyrate is preferentially utilised by the cells that line the large intestine, while more acetate and propionate find their way into the general blood circulation. Propionate is mostly metabolised in the liver. This leaves acetate as the short chain fatty acid that reaches the highest concentrations in the peripheral blood that supplies most tissues. Acetate gives rise to energy as it can go straight into the TCA cycle after being converted to acetyl-CoA (Chap. 4). Butyrate is converted to two acetyl-CoA, while propionate (a fatty acid with an odd number of carbons) is used after being converted first to glucose.

Most of these SCFA come from microbial fermentation of dietary material that is not digested by the host itself (i.e. fibre). Therefore, the microbiota must clearly be contributing additional calories via SCFA that would not be available to the host in the absence of gut micro-organisms. As discussed already, this additional energy is

© Springer Nature Switzerland AG 2020
H. J. Flint, *Why Gut Microbes Matter*, Fascinating Life Sciences,
https://doi.org/10.1007/978-3-030-43246-1_8

crucial for the survival of animals that rely on diets high in plant fibre and SCFA can account for 80% of dietary energy gain in ruminants [1]. SCFA may also be crucial energy sources in humans (both now and through our prehistory) who subsist on diets containing substantial amounts of fibre. Many individuals in modern societies however suffer from the opposite problem, as we recognise that overconsumption of calories and excess energy gain is contributing to the rapidly increasing incidence of overweight and obesity. This raises the question, to what extent might the fermentative activities of our gut microbiota be contributing to the problem of obesity?

Microbial Fermentation and Obesity

How significant is the 'additional energy' (which has also been termed *energy harvest* [2]) gained from microbial fermentation of dietary fibre in humans? At first sight, germ-free animals (that are artificially delivered and raised so as to have sterile intestines) appear to provide a brilliant way to determine whether gut microorganisms might contribute to obesity. Some studies have found that germ-free mice put on less weight and less fat than do *conventional* animals (i.e. those with their normal rodent gut microbiota) when fed a high fat ('Western style') diet [3]. Is this because they are harvesting less energy from the diet? Another study using a different high fat diet, however, has reported precisely the opposite outcome, with germ-free animals gaining more weight than conventional animals [4]. It is suggested that this may reflect differences in the type of fat and the content of fibre and sucrose between the 'high fat' diets, although other differences such as in physical activity and energy expenditure might also have played a role [4–6]. Therefore, although the germ-free mouse model is widely cited, its interpretation is far from straight-forward.

While it is possible to study germ-free rodents that have no gut micro-organisms (and have to be maintained in isolators in the laboratory) this is not an option either ethically or practically for humans. We do not have a choice as to whether or not we are colonised by gut microbiota. For us, the relevant question to ask is whether certain configurations of our gut microbiota yield more energy via SCFA production than others. If this is so, might it be that obese people carry microbiota configurations that tend to promote greater recovery of energy via SCFA when compared to lean people? Exactly this proposal has been indeed been made in a paper suggesting that obese humans possess a higher proportion of Firmicutes and a lower proportion of Bacteroidetes bacteria in their faecal microbiota compared with lean individuals [7]. Separate evidence from a rodent study indicated that energy harvest in normal lean mice was less efficient than in genetically obese mice, and that the gut microbiota of the obese mice showed a higher proportion of Firmicutes relative to Bacteroidetes compared with the lean mice [2]. On the other hand, studies in human volunteers that used fluorescent probes to directly enumerate the faecal microbiota have shown that microbiota composition changes little at the phylum level, or tends to slightly higher % Bacteroidetes, with increasing obesity (as measured by body mass index, or BMI) [8, 9]. Subsequent large metagenome-based studies have found some differences in faecal microbiota profiles between obese and lean individuals, but these are mostly at the level of individual species rather than bacterial phyla [10].

It should not be a major surprise to find that faecal microbiota profiles change with increasing BMI. This is because we can reasonably assume that eating behaviour, and therefore dietary intake, contributes to the obese state, and we know that diet (especially fibre intake) is a major driver of changes in the microbiota composition, as was discussed in Chap. 6. Microbiota changes might therefore be expected to accompany the dietary habits that lead to weight gain. The key question, to which we must return later, is whether such alterations in gut microbiota profiles are actually contributing directly to the obese state? The alternative (if less exciting) explanation is simply that dietary intakes that lead to obesity also promote a change in gut microbiota profiles, with no causal link necessarily involved.

We have seen that the initial hypothesis that differences in energy harvest resulting from variation in gut microbiota composition are a major factor in obesity has received little support from subsequent studies [11]. But we did note earlier (in Chap. 6) one species difference within the human gut microbiota that must have resulted in altered energy harvest from the diet. In most people microbial fermentation of resistant starch (RS) is close to 100%, but in one of our own studies we found there were two individuals whose gut microbiota lacked the starch-degrading species *Ruminococcus bromii*. These two individuals were the only ones in the study to show much residual starch in their stool samples and their gut microbiota fermented only about half of the RS from their diet [12]. The key question is then, how much effect could this have had on the host's overall energy balance?

Energy from Dietary Fibre

Definitions of fibre (fiber) both in the EU and USA now include all non-digestible soluble and insoluble carbohydrates with three or more sugar monomers. As such, fibre includes resistant starch as well as soluble prebiotic fibres, in addition to non-starch polysaccharides. Adult human fibre intake in the UK is estimated to be around 17–20 g/day [13] and in the USA around 16 g/day [14].

Digestible carbohydrates are estimated to contribute 3.8 kcal/g to our supply of energy through metabolism of glucose in body tissues.[1] The contribution of energy from carbohydrates in fibre is harder to calculate precisely but must inevitably be less than this (Fig. 8.1). In part, this is because gut bacteria use up some (around a quarter) of the energy derived from the sugars that they release from fibre for their

---

[1] 'kcal' (kilocalories) is the proper term, but is commonly abbreviated to 'calories'. Slightly different values can be found in the literature for the calorific values of starch, sucrose and glucose and a value of 4 kcal/g (17 kJ/g) is commonly used in food labelling. The figure given here is from the British Nutrition Foundation [13].

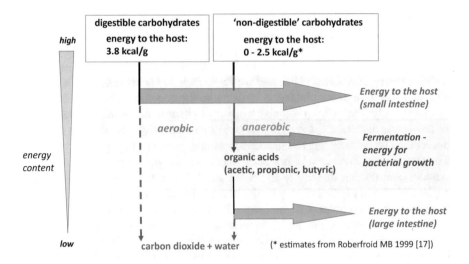

Fig. 8.1 Contribution of ingested carbohydrates to dietary energy supply to the host

own growth. In addition, fermentation of the fibre by the micro-organisms, and absorption of the resulting SCFA by the gut, are not 100% efficient, which results in some of the calories being lost in the faeces. While SCFA absorption is highly efficient (up to 95%), perhaps only 50–75% of the fibre arriving in the large intestine gets fermented [15]. We also know that the recovery of energy from fibre depends on how fast digesta pass through the intestine (gut transit), with both fermentation and SCFA absorption becoming less complete with more rapid passage [16]. An approximate figure of 2 kcal/g fibre gained via microbial SCFA is at the top end of a range of estimates that vary with the type of fibre in the diet [17].

If we could measure exactly how much SCFA were being absorbed from the gut, we could then estimate their energy contribution directly. This is very difficult to do in humans, but precise measurements of microbial SCFA production have been made in the gut of pigs, where SCFA were found to contribute around 10% of the animal's energy requirements [18]. More recent estimates from pigs for the contribution of fermentation to energy requirements are 10.7% for diets with 75 g fibre/kg feed, increasing to 24% on a diet extremely high in fibre (147 g/kg) [19]. As we saw in Chap. 2, however, pigs have a much greater fermentative capacity than humans.

An alternative, indirect, approach to estimating how much human diet-derived energy comes via fermentation in humans has been to ask how much energy is needed to account for the mass of bacteria present in the gut, and then to deduce how much total fibre would have to be fermented to produce this much energy. This line of reasoning led to a proposal that, in addition to non-starch fibre, 10–15% of dietary starch and sugar (corresponding to 35–50 g/day) from 'Western' diets reaches the large intestine [15]. These figures are hard to verify as the amounts of non-digested starch and sugars are particularly difficult to measure directly. This calculation is however the basis for a frequently cited estimate of 10% for our daily energy gain coming from non-digestible carbohydrates [15]. A major problem with these calculations is that, for reasons explained in Chap. 3, we simply do not know precisely how much ATP is released by carbohydrate fermentation by anaerobic bacteria, or how much ATP is required to produce a gram of bacterial cells, in the gut. Indeed, this will vary with the composition of the microbiota. This is an issue where confident assertions can often be found, but these turn out to be based on very few primary sources.

Some Digestive Arithmetic

For the sake of argument, let us consider 0 g, 25 g, 50 g and 100 g as alternative total daily fibre intakes in humans (equivalent to 0, 55, 110 and 220 g fibre/kg total intake in the diets considered here). As we have seen, 50 g is higher than current recommendations while 25 g is closer to the average adult fibre intakes in the UK and USA. The 100 g fibre intake is almost certainly unachievable and is included for theoretical comparison only. On this basis, some simple arithmetic allows us to estimate the impact of 'energy harvest' from fibre on total dietary calories. A total fibre intake of 25 g/day compares, typically, with a more than 10 times higher daily intake of digestible (non-fibre) carbohydrates. Taking all the calories that come from digestible fibre, protein and fat in our typical diet into account, we would predict that around 2.5% of total energy would then come from the fibre (assuming fibre contributes 2 kcal/g) (Fig. 8.2). If the efficiency of fermentation was halved, say because of an 'inefficient' microbiota composition, this would lead to a drop of around 25 kcal in total energy harvest—or the equivalent of half a biscuit a day. Such small changes can of course affect weight gain over a long period of time, and this would correspond to a little over 1 kg body weight over a year. It is worth noting that the effect will be smaller still in individuals who normally consume a diet that is lower in fibre but greater in those with higher fibre intakes. We should note that many rural communities around the world may have greater total fibre intakes and therefore a higher % contribution of calories from microbial fermentation [4]. Mean intakes of 28 g plant NSP fibre and 38 g RS per day were recorded for rural Africans in one recent study [20].

It is worth looking more closely at what happens when we replace some of the digestible carbohydrate in our diet with fibre. The answer (Fig. 8.2) is that with a 50 g fibre intake we gain a higher proportion of our calories (5%) from microbial fermentation (SCFA), but the total energy gain from the diet has actually decreased slightly. If the fibre intake were 100 g we might then gain 10% of calories from fermentation, but with a further decrease in total energy gain. Given that the SCFA themselves appear to confer additional health benefits (as will be discussed later) this suggests that the increase in dietary fibre content and in fermentation actually provide a net benefit with regard to weight control and health maintenance. This is in line with new dietary guidelines that have increased the recommended daily fibre intake.

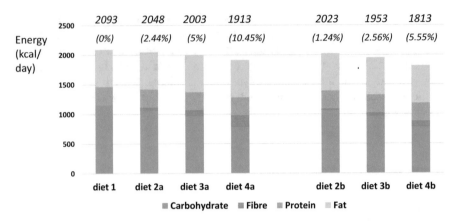

Fig. 8.2 Predicted effects of energy harvest from dietary fibre in humans. Including fibre in place of digestible carbohydrate (1, 2a, 3a, 4a) is predicted to decrease overall energy supply from the diet. Decreasing the efficiency of microbial fermentation would also reduce energy supply (2b, 3b, 4b compared to 2a, 3a, 4a). Total kcal are shown above the bars with % coming from the fermentation of fibre in parentheses. All four diets assume 455 g/day total food intake, including 80 g fat and 70 g protein. Fibre includes non-starch fibre and RS. *Diet 1* assumes 305 g non-fibre carbohydrate (CHO) and no fibre. *Diet 2a* assumes 280 g non-fibre CHO and 25 g fibre intake; *Diet 2b* the same, but 50% fermentation efficiency. *Diet 3a* assumes 255 g non-fibre CHO and 50 g fibre intake; *Diet 3b* the same, but 50% fermentation efficiency. *Diet 4a* assumes 205 g non-fibre CHO and 100 g fibre intake; *Diet 4b* the same, but 50% fermentation efficiency

Most of us are familiar with the benefits of fibre intake in promoting regular toilet habits, as a result of speeding up passage through the intestine (increasing the rate of gut transit). In general, faster passage of digesta through the intestine is predicted to lead to less complete digestion and fermentation of food components, but also to higher microbial cell populations and SCFA concentrations and lower gut pH. Indeed, these effects have been demonstrated in human volunteers directly using agents that speed up or slow down gut transit [16]. We found recently that inclusion of different fibres in high fat diets helped to limit weight gain in conventional mice. But while these fibres had very different effects on caecal microbiota composition, their effects on weight gain were similar, suggesting a common mechanism [21].

So, to summarise, we appear to have come full circle! Microbial fermentation of fibre does indeed provide additional calories that we would not otherwise be able to access from our diets, thus increasing 'energy harvest' from the diet. Except for those on very high fibre intakes, however, the contribution to total calorific intake is relatively small. The more important point is that for a constant intake of food, a higher proportion of fibre compared with digestible carbohydrate in our diet actually leads to a net decrease in total energy gain from the diet. This is because of the lower net energy yield per gram from fibre to the host (ourselves) together with possible effects of fibre in promoting *satiety*.

How Else Might the Microbiota Influence Weight Gain and Metabolism?

Does this mean that gut microbes are likely to have little or no impact on obesity? Well—not necessarily. There is substantial body of evidence (mainly from studies with animal models, discussed below) to suggest that they may indeed have a role. We also have to recognize that gut microbes may not exert their effects solely, or even mainly, through the yield of energy from fermentation [6]. A variety of other mechanisms are now seen as possibilities. Microbial activity might somehow influence how hungry we feel (appetite and *satiety*) and so affect how much we eat. This is indeed a real possibility, as there is evidence that microbially-produced SCFA, especially propionate, affect the production of particular hormones that control satiety and make us feel less hungry [22]. Conversely, gut micro-organisms might influence how much energy we use, either by affecting levels of physical activity or by imposing an additional burden on the host. The latter undoubtedly applies in the case of gut infections, which require extra energy to help restore the integrity of the gut barrier and to enable the immune system to combat the infection.

Gut Microbiota and Obesity in Small Animal 'Models'

The gut of germ-free animals can be colonised experimentally with mixed human faecal bacteria, single bacterial strains or with defined mixtures of micro-organisms. These are known as gnotobiotic (for 'known microbiota') animals. One intriguing study examined initially gnotobiotic mice whose gut had been re-colonised with a defined mixture of 10 bacterial species [23]. It was shown that missing out one of these species (*Clostridium ramosum*) from the mixture resulted in reduced fat deposition and weight gain, which suggests that this bacterium was somehow promoting obesity. Conversely, other researchers found evidence that the gut bacterium *Christensenella* promoted reduced weight gain when introduced into germ-free mice. Their experiments were prompted by observations from twin studies that *Christensenella* tends to show higher abundance in lean compared with obese humans [24]. Such experiments with gnotobiotic animals do suggest that there is potential for some members of the gut microbiota to influence obesity. The mechanisms have still to be clarified, although *C. ramosum* is reported to stimulate fat absorption [25]. The significance of these findings relative to other factors that promote human obesity remains hard to assess, however, particularly since neither of the above-mentioned species is particularly abundant within the normal human gut microbiota.

Antibiotics and Obesity

In farm animals, antibiotics have been used for decades as *growth promoters*. In most European countries legislation has been passed to restrict such usage to antibiotics that are thought not to have therapeutic value for treating human diseases. This is because of very real concerns about promoting the spread of antibiotic resistance [26]. Elsewhere (in much of the world, including the USA) there are far fewer restrictions on antibiotic usage in farming.

But why should antibiotics that tend to kill, or slow the growth, of gut bacteria promote better growth of the host animal? In fact, the mechanisms appear to differ according to the species of farm animal. In ruminants, ionophore antibiotics such as monensin (which were first approved to control coccidial parasites) have been widely used. These appear to alter the rumen fermentation by suppressing hydrogen-producing bacteria and hence methane formation, probably helping to deliver more energy to the animals via short chain fatty acids. This is assumed to improve 'energy harvest' and production efficiency (see Chap. 4). On the other hand, in pigs, whose gut anatomy is much closer that of humans, antibiotics have been used mainly with the aim of combatting bacterial infections that tend to occur around the time of weaning. This allows earlier weaning and hence more rapid weight gain than would otherwise be possible. The antibiotic-treated animal also avoids the huge energy and other costs involved in fighting off infections.

Antibiotics are used in human infants to prevent and treat what could otherwise be damaging or life-threatening infections. There is a trend towards increased antibiotic usage in early life along with increased frequency of delivery of babies by caesarian section, particularly (but not only) in the USA. It has been argued from the results of animal studies that this early use of antibiotics may be contributing to obesity in later life as a result of imbalances caused in the developing gut microbiota [27]. Intriguingly, a survey of Danish mothers and children found statistical evidence to suggest that antibiotics administered within the first 6 months of life promoted overweight among children born to normal-weight mothers but decreased overweight among children born to obese mothers [28]. This might possibly be interpreted to imply that antibiotics inhibit 'lean-promoting' species in the microbiota of children of normal weight mothers but inhibit 'obesity-promoting' species in the microbiota of children of obese mothers. On the other hand, another study of over 250,000 children found no evidence that early antibiotic administration leads to childhood obesity, although infections not treated with antibiotics were associated with subsequent overweight [29]. It appears that there is currently no clear evidence that early antibiotic use leads to obesity in humans.

Gut Microbiota, Metabolic Health and Diabetes

The concentration of glucose in our blood is controlled mainly by two peptide *hormones* that are produced by the pancreas—insulin and glucagon. Insulin increases the removal of glucose from the bloodstream, promoting its transport into tissues to provide an immediate source of energy and also its conversion into the starch-like polysaccharide glycogen or into fats. These latter compounds act as stores of energy for later use that are held in muscle, liver or adipose tissues (fat cells). Glucagon does the opposite of insulin, promoting the breakdown of glycogen to produce glucose. In the healthy state, the formation of both of these hormones is controlled by the blood glucose concentration. Higher blood glucose leads to more insulin and less glucagon being produced, and lower blood glucose to less insulin and more glucagon, with the net result being tight control over the level of the glucose in the blood (Fig. 8.3). Failure of this finely balanced system is the cause of diabetes, in which blood glucose increases and has to be excreted in the urine while the tissues themselves are starved of glucose. In type 1 diabetes, insufficient insulin is produced because of damage to the insulin-secreting cells of the pancreas. Successful treatment involves injections of insulin to restore glucose control. The more common condition type 2 diabetes, however, involves in addition the progressive failure of the body to respond to the insulin that is produced. This phenomenon is known as 'insulin resistance' and is an increasingly common consequence of poor diets and overweight in the human population. The term 'metabolic syndrome' is often used for a collection of symptoms (including high blood pressure, elevated blood glucose and increased waist circumference) that may indicate a progression towards type 2 diabetes.

The glucose in our bloodstream in fact comes from two distinct sources. The one that we have mainly discussed so far is glucose that is derived from the breakdown of carbohydrate chains, arising either from the digestion of food in the gut or from

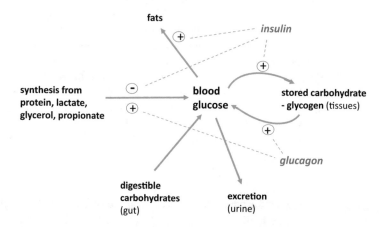

Fig. 8.3 Control of blood glucose by the hormones insulin and glucagon

glycogen stored in body tissues. A second important route however involves the new synthesis of glucose (*gluconeogenesis*) from pyruvate that is derived from particular amino acids, or from other non-carbohydrate sources including lactate, propionate and glycerol. This allows the supply of energy to body tissues to be maintained when there is little glucose coming from the diet. This process of new synthesis is also controlled by insulin. Diabetics have higher rates of new glucose formation than healthy controls, which contributes to their high blood glucose concentrations and the consequent loss of glucose in the urine. The drug metformin that is commonly used to treat type 2 diabetes acts largely by limiting new glucose formation by the liver [30].

Several human studies using metagenomic techniques have detected differences in faecal microbiota profiles between type 2 diabetic and control subjects. Most of these studies have not considered the possible effects of the drugs used to treat type 2 diabetics however, and it turns out that taking metformin itself results in altered microbiota profiles [31]. Nevertheless, type 2 diabetics who are not taking metformin still show changes in their gut microbiome when compared with healthy controls. In particular, gut bacteria known to produce butyrate are among those found to be decreased in type 2 diabetics [31].

The activities of the gut microbiota might be linked to type 2 diabetes in a number of ways. As will be discussed in Chap. 9, microbially-produced short chain fatty acids (SCFA) appear to help regulate food intake, glucose supply and metabolism. Indeed, there is evidence in mice that propionate and butyrate, acting via different mechanisms, promote gluconeogenesis in intestinal cells, which results in beneficial changes in systemic glucose metabolism [32]. Inflammation is an important contributor to metabolic syndrome and type 2 diabetes, so that the established anti-inflammatory effects of SCFA may also be protective. This means that in general the microbial fermentation of fibre may be helpful in preventing metabolic disease. Conversely, lipo-polysaccharide (LPS) associated with Gram-negative bacteria (which is also referred to as an endotoxin) is a potent cause of inflammation and there is evidence that LPS in the bloodstream, arising from infection or leakage from the gut, can promote the symptoms of diabetes [33]. Diet-derived compounds such as long chain fats can also promote inflammation. It remains unclear however whether LPS is a major factor in impaired glucose metabolism in diabetes [34]. Some elements of the gut microbiota, such as bifidobacteria, are thought to help promote integrity of the *gut barrier* and so prevent 'leakiness' and this is a possible mechanism through which probiotics and prebiotics could help to prevent inflammation [35]. Some success has also been claimed for faecal microbiota transfer (FMT) in type 2 diabetes [36]. In this procedure, faecal bacteria from a healthy 'donor' are introduced into the patient's large intestine (discussed further in Chap. 12). One the bacteria that was most prominent in the restored, healthy microbiota was *Anaerobutyricum hallii*, which has a special role in organic acid metabolism through conversion of lactate and acetate to butyrate. All this being said, the major treatment regimes for type 2 diabetes continue to be through diet, weight loss, exercise and medication aimed at controlling glucose formation, insulin production and insulin sensitivity.

Conclusions

Obesity and its associated health problems, which include type 2 diabetes, heart disease and several types of cancer, are among the biggest health problems confronting modern societies. This applies not only in the Western world, but increasingly also in Asia and elsewhere as traditional dietary patterns shift towards less healthy food intakes. Following some striking early observations, the possible involvement of the gut microbiota in obesity and type 2 diabetes has attracted much interest and indeed excitement among researchers. At this point, however, we have to be somewhat cautious and say that there is still relatively little hard evidence to establish that there are microbially-mediated mechanisms leading directly to the development of human obesity. On the other hand, it does appear that certain microbiota profiles are associated with unhealthy diets and may be a factor in the negative effects of these diets. This suggests that that we need to focus more specifically on how microbial activity may promote, or protect against, major health problems in which diet plays an important role, notably colorectal cancer, type 2 diabetes and cardiovascular disease.

References

1. Bergman EN (1990) Energy contribution of volatile fatty acids from the gastrointestinal tract in various species. Physiol Rev 70:567–590
2. Turnbaugh PJ et al (2006) An obesity-associated gut microbiome with an increased capacity for energy harvest. Nature 444:1027–1031
3. Backhed F et al (2004) The gut microbiota as an environmental factor that regulates fat storage. Proc Natl Acad Sci USA 101:15718–15723
4. Fleissner CK et al (2010) Absence of intestinal microbiota does not protect mice from diet-induced obesity. Br J Nutr 104:919–929
5. Arora T, Backhed F (2016) The gut microbiota and metabolic disease: current understanding and future perspectives. J Intern Med 280:339–349
6. Flint HJ (2011) Obesity and the gut microbiota. J Clin Gastroenterol 45(Suppl):S128–S132
7. Ley RE et al (2006) Microbial ecology: human gut microbes associated with obesity. Nature 444:1022–1023
8. Duncan SH et al (2008) Human colonic microbiota associated with diet obesity and weight loss. Int J Obes 32:1729–1724
9. Schwiertz AD et al (2010) Microbiota and SCFA in lean and overweight healthy subjects. Obesity 18:190–195
10. Le Chatelier E et al (2013) Richness of human gut microbiome correlates with metabolic markers. Nature 2013(500):541–546
11. Murphy EF et al (2010) Composition and energy harvesting capacity of the gut microbiota: relationship to diet, obesity and time in mouse models. Gut 59:1635–1642
12. Walker AW et al (2011) Dominant and diet-responsive groups of bacteria within the human colonic microbiota. ISME J 5:220–230
13. British Nutrition Foundation (2318) Dietary fibre. https://www.nutrition.org.uk/nutrition science/nutrients-food and ingredients/dietary-fibre.html

14. McGill CR et al (2015) Ten-year trends in fiber and whole grain intakes and food sources for the United States population: national health and nutrition examination survey. Nutrients 7:1119–1130
15. McLean NI (1984) The contribution of the large intestine to energy supplies in man. Am J Clin Nutr 39:338–342
16. Stephen AM et al (1987) Effect if changing transit time on colonic microbial metabolism in man. Gut 28:601
17. Roberfroid MB (1999) Calorific value of inulin and oligofructose. J Nutr 129:1436S–1437S
18. Imoto S, Namioka S (1978) VFA production in the pig intestine. J Anim Sci 47:467–478
19. Iyayi EA, Adeaola O (2015) Quantification of short chain fatty acids and energy production from hindgut fermentation in cannulated pigs fed graded levels of wheat bran. J Anim Sci 93:4781–4787
20. O'Keefe SJ et al (2015) Fat, fibre and cancer risk in African Americans and rural Africans. Nat Commun 6:6342
21. Drew JE et al (2018) Dietary fibers inhibit obesity in mice, but host responses in the cecum and liver appear unrelated to fiber-specific changes in cecal bacterial taxonomic composition. Sci Rep 8:15566
22. Morrison DJ, Preston T (2016) Formation of short chain fatty acids by the gut microbiota and their impact in human metabolism. Gut Microbes 7:189–200
23. Woting A et al (2014) *Clostridium ramosum* promotes high-fat diet-induced obesity in gnoto-biotic mouse models. MBio 5:1530-14
24. Goodrich JK et al (2014) Human genetics shape the gut microbiome. Cell 159:789–799
25. Mandic AD et al (2019) *Clostridium ramosum* regulates enterochromaffin cell development and serotonin release. Sci Rep 9:1177
26. Davies J, Davies D (2010) Origins and evolution of antibiotic resistance. Microbiol Mol Biol Rev 74:417–433
27. Cho I et al (2012) Antibiotics in early life alter the murine microbiota and adiposity. Nature 488:621–626
28. Ajslev TA et al (2011) Childhood overweight after establishment of the gut microbiota: the role of delivery mode, pre-pregnancy weight and early administration of antibiotics. Int J Obes 35:522–529
29. Li D-K et al (2017) Infection and antibiotic use in infancy and risk of childhood obesity: a longitudinal birth cohort study. Lancet Diabetes Endocrinol 5:18–25
30. Hundai RS et al (2000) Mechanisms by which metformin reduces glucose production in type 2 diabetics. Diabetes 49:2063–2069
31. Forslund K et al (2015) Disentangling type 2 diabetes and metformin signatures in the human gut microbiota. Nature 528:262–266
32. De Vadder F et al (2014) Microbiota-generated metabolites promote metabolic benefits via gut-brain neural circuits. Cell 156:84–96
33. Cani PD et al (2007) Metabolic endotoxemia initiates obesity and insulin resistance. Diabetes 56:1761–1772
34. Caesar R et al (2012) Gut-derived lipopolysaccharide augments macrophage accumulation but is not essential for impaired glucose or insulin tolerance in mice. Gut 61:1701–1707
35. Cani PD et al (2007) Selective increases in bifidobacteria in gut microflora improve high-fat diet-induced diabetes in mice through a mechanism associated with endotoxemia. Diabetologia 50:2374–2383
36. Vriese A et al (2012) Transfer of intestinal microbiota from lean donors increases insulin sensitivity in individuals with metabolic syndrome. Gastroenterology 143:e7

Chapter 9
Gut Microbes and Metabolites

We are largely insulated from direct contact with the cells of our large intestinal microbiota by the thick layer of mucin that coats the healthy intestinal wall. On the other hand, the small molecules (metabolites) that are produced by microorganisms can diffuse through the mucin layer. As a result, the biochemical activities of our gut microbiota have a major impact on our physiology and health. Microbial metabolites not only determine the environment at the gut surface, but many of them are taken up into the bloodstream [1]. Here we will look briefly at some of the metabolic activities most relevant to health.

Microbial Production of Short Chain Fatty Acids

Microbial fermentation gives rise to a range of fatty acids from one to six carbons in length (thus referred to as 'short chain fatty acids' or SCFA). Three of these (acetic, propionic and butyric acids) typically account for around 90% of the total SCFA in colon contents and are found in the approximate ratio of 3 acetate: 1 propionate: 1 butyrate in faecal samples [2]. These acids arise mainly from the fermentation of dietary fibre (Chap. 7) although they are also produced from the fermentation of amino acids in the gut. In addition, some minor SCFA (isobutyrate, 2-methyl butyrate, isovalerate) arise only from the fermentation of branched chain amino acids. The proportions in which these various SCFA are produced in the gut can be affected by microbiota composition, by gut pH, and potentially also by the substrates being fermented [3, 4].

Overall SCFA concentrations in the colon are generally high (70–180 mM)[1] which makes them the most important source of negatively charged ions (anions)

---

[1]mM = millimolar. A molar concentration of any compound is defined as the molecular weight of that compound in grams dissolved in a litre of water. For example, one rather well-known soft drink contains 106 g sugar per litre volume. Since the molecular weight of sucrose is 342, a molar solution would contain 342 g/L and the soft drink is 0.311 molar (M) or 311 millimolar (mM) sucrose.

© Springer Nature Switzerland AG 2020
H. J. Flint, *Why Gut Microbes Matter*, Fascinating Life Sciences,
https://doi.org/10.1007/978-3-030-43246-1_9

in gut contents. Transport of SCFA across the gut wall involves exchange with bicarbonate anions, which helps to control gut pH, and transportation in parallel with positively charged sodium (Na^+) and hydrogen (H^+) ions [5]. Consequently, SCFA play an important role in the uptake of salt and water from the large intestine. As discussed already, SCFA also provide the energy that the host gains from microbial fermentation. Some 95% of the SCFA produced within the gut are absorbed across the gut wall [2]. Butyrate concentrations in the blood circulation are relatively low because of the efficient use of butyrate by gut epithelial cells, but propionate and acetate reach higher concentrations in the blood supply that leaves the gut wall.

SCFA have multiple effects on human cells and tissues, most of which are beneficial to health. Butyrate is the preferred source of energy for the cells that line the large intestine, but it also changes the expression of many genes in these cells. This is partly because it inhibits an enzyme (histone deacetylase) that controls the unfolding of host chromosomes and hence the availability of their genes for transcription. In addition, butyrate is recognised by molecules known as *receptors* (FFAR or free fatty acid receptors) that reside on the surface of many types of human cell. These multiple molecular effects of butyrate have important consequences for our physiology and health that include the promotion of cell death (*apoptosis*) in colorectal cancer cells and the suppression of inflammation. The cells that line that large intestine (epithelial cells) have evolved to use butyrate, rather than glucose, as their preferred source of energy. These cells become energy starved when the supply of butyrate is limited, for example when its transport is inhibited by sulfide, making butyrate a key factor in maintaining a healthy gut. Utilization of butyrate by host epithelial cells involves a process (β-oxidation) that uses oxygen, limiting oxygen diffusion into the gut and helping to maintain the anaerobic conditions that favour obligate anaerobes [6]. Limitation of butyrate supply, however, causes these cells instead to ferment glucose to lactate and increases the release of oxygen into the gut—conditions that can promote 'dysbiosis' and favour the growth of facultatively anaerobic pathogens.

Like butyrate, propionate also inhibits histone deacetylase, while both propionate and acetate also interact with FFARs. The FFARs that respond to microbially-produced SCFA (FFAR2 and FFAR3) are different from those recognised by long and medium chain fatty acids that are acquired directly from the diet, and their stimulation has different physiological consequences (Table 9.1). FFAR2 and FFAR3 are found in the intestinal wall, peripheral nervous system, cells of the innate immune system, adipose (fat) tissue and pancreatic cells. Their stimulation influences cellular processes that include inflammation, glucose metabolism, sensitivity to insulin and the secretion of hormones [7]. One consequence of FFAR stimulation by propionate is the increased production of hormones that help to prevent us feeling hungry (i.e. that promote satiety) [8]. Propionate also helps to decrease cholesterol formation by inhibiting one of the enzymes involved in

Table 9.1 Binding affinities of human free fatty acid receptors for fatty acids

	FFAR3 (GPR41)	FFAR2 (GPR43)	FFAR1 (GPR40)	FFAR4 (GPR120)	
Saturated					
Acetic (C2:0)	+	+++	–	–	
Propionic (C3:0)	++++	+++	–	–	*Short chain*
Butyric (C4:0)	+++	+++	–	–	*fatty acids*
Valeric (C5:0)	+++	+	–	–	*(SCFA)*
Caproic (C6:0)	++	–	–	–	
Palmitic (C16:0)	–	–	++++	+++	
Stearic (C18:0)	–	–	+++	++++	
Mono-unsaturated					
Oleic (C18:1)	–	–	++++	+++	
Polyunsaturated (ω-3)					
α-linolenic (C18:3 n-3)	–	–	++++	+++++	
Polyunsaturated (ω-6)					
Linoleic (C18:2 n-6)	–	–	++++	+++++	

Table simplified from Miyamoto et al. 2016 [7] which gives numerical binding affinities

cholesterol synthesis. Acetate and propionate have further important beneficial effects in decreasing glucose formation in the liver and in controlling the formation and breakdown of body fat. Some of these effects are thought to involve increases in the activity of a key host enzyme (AMPK) that is involved in regulating these processes [5].

To summarise, the SCFA produced by anaerobic gut bacteria may account for many of the beneficial health effects that we gain from consuming fermentable fibre in our diet. Butyrate appears to have a particularly important role in gut health as it helps to protect against colorectal cancer and colitis. Propionate and acetate appear more protective for metabolic health, through control of cholesterol and glucose metabolism.

Phytochemicals and Drugs

Plants produce a huge range of exotic chemical compounds that are well beyond the capabilities of our own enzyme systems to manufacture. These compounds, many of which are based on carbon ring (or 'aromatic') structures, are often loosely termed *phytochemicals*. They include some potent toxins (e.g. atropine, strychnine) and drugs (e.g. morphine, caffeine) but most of them are harmless, or offer some health benefits. Indeed, the familiar aspirin is a modified form of the phytochemical compound salicylic acid, which occurs widely in plants.

Fig. 9.1 Enterohepatic circulation of phenolic compounds. In the liver, phenolic compounds (phe) are conjugated, here with glucuronic acid, and released into the gut. Microbial glucuronidases release the phenolic group, which may then return to the liver. Phenolic compounds linked to sugars in food are released by gut microbial glycosidases

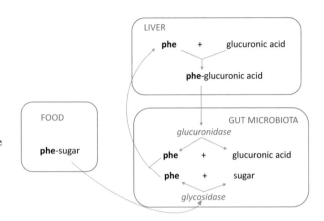

Phytochemicals do not generally exist on their own within food plants. They may be bound (conjugated) to sugars, or they may be locked into complex structures such as plant cell walls. As a result, many of them pass into the large intestine without being absorbed in the upper gut, and so become exposed to the resident gut microbiota. Microbial activity results in the release of more phytochemicals as the plant material is degraded, but it can also alter them to create further compounds that do not exist in the plant [9]. An example is ferulic acid, which is associated with plant lignin and is released only when plant cell wall structures are broken down by specialist bacteria in the colon [10]. Other bacteria within the gut microbiota then modify ferulic acid to yield other phenolic acids. Ferulic acid and its metabolites have the potential, once absorbed by the body, to limit oxidative damage to human cells. Compounds such as ferulic acid, which is just one of a great many *bioactive* phytochemical compounds, are thought to contribute significantly to the health protective effects of plant fibre and 'whole grain' products.

Once in the bloodstream these foreign plant compounds (often called *xenobiotics*) pass through the liver, where they are joined (*conjugated*) to other molecules (most commonly, glucuronic acid or sulfate) as a prelude to their disposal in the bile or in urine. This is part of the liver's crucial role in detoxifying the blood. The release of these liver conjugates back into the gut in the bile, however, exposes them again to the gut microbiota. Unsurprisingly, perhaps, there are some gut micro-organisms that can break (*deconjugate*) the linkages involved in the conjugates, so releasing the free phytochemical which may then find its way back into the bloodstream and to the liver. Bacterial β-glucuronidase enzymes are particularly important in this activity (Fig. 9.1).

The upshot of all this is an *enterohepatic circulation* of phytochemical compounds which shuttle between being conjugated in the liver and being de-conjugated in the gut. This phenomenon is not merely of academic interest to biochemists and physiologists. Many common medicinal drugs (such as paracetamol (or Tylenol)) are subject to this complex circulation. What this means is that doses required for effective treatment may vary according to the composition and activities of an individual's gut microbiota. This complicates decisions on the correct dosage to use and the prediction of treatment outcomes.

Bile Acids

Cholesterol is an important lipid (fat) in the human body, as a major component of cell membranes and the source of many steroid hormones. Excess cholesterol has acquired a bad reputation, however, because of its association with heart disease. Enzymes in the liver convert cholesterol to bile acids, which are secreted into the intestine from the gall bladder in bile thus helping the body to dispose of excess cholesterol. Once secreted, bile acids play an important digestive role by emulsifying fats. More cholesterol is formed when the diet is high in saturated fats and consequently the secretion of bile and of bile acids increases when we are consuming such diets.

The bile acids made by the liver are known as primary bile acids. In humans there two predominant types, cholic acid and chenodeoxycholic acid. Also within the liver, and before they are secreted, these bile acids are joined (conjugated) to one of two amino acids. The amino acids concerned are glycine (the smallest of the 20 amino acids found in proteins) and taurine, a sulfur-containing amino acid that is not found in proteins. Taurine is made from the sulfur-containing amino acids methionine and cysteine but is not itself broken down by the body and has to be disposed of either in urine or in bile secretions. Interestingly, it is the nature of the diet that largely dictates whether the bile acids are conjugated to glycine (as happens with largely plant-based diets) or to taurine (as happens with diets high in animal protein and fat). This is likely to have major consequences for health, mainly because of the action of intestinal bacteria on these compounds.

When these conjugated bile acids reach the large intestine, several resident bacterial species produce an enzyme (bile salt hydrolase) that breaks the linkage that joins the bile acids to the amino acids. In other words, they hydrolyze the conjugates formed in the liver (i.e. they are responsible for de-conjugation). In the case of the taurine conjugate taurocholic acid, this de-conjugation releases taurine and choline into the intestine (Fig. 9.2). This now presents a wonderful opportunity for certain specialised members of the gut microbiota. In particular, the bacterium *Bilophila wadsworthii* is able to grow by breaking the taurine down to release ammonia, acetate and hydrogen sulfide (H_2S). When the host cells that line the gut are exposed to high concentrations of H_2S, it can damage cellular DNA, cause inflammation and block the utilization of butyrate as an energy supply. The other compound resulting from de-conjugation is the primary bile acid cholic acid. Again, there are bacteria (for example *Clostridium scindens*) that specialise in acting on cholic acid, removing a hydroxyl group to convert it into the *secondary bile acid* deoxycholic acid DCA. Unfortunately, DCA (unlike cholic acid) is a tumour-promoting agent. In summary, the combination of genotoxic H_2S and tumour

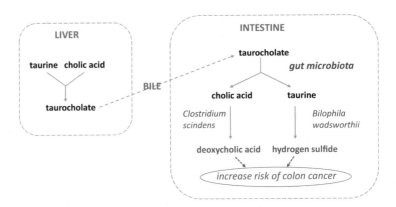

Fig. 9.2 How microbial metabolism of bile acids may increase the risk of colon cancer. On diets high in protein, cholic acid is conjugated mainly with taurine in the liver, whereas on 'plant-based' diets it is conjugated mainly with glycine (see text) (Ridlon et al. 2016 [11])

promotion resulting from microbial metabolism of taurocholate contributes to an environment that is likely to promote colorectal cancer [11].

Rural native Africans who consume diets high in resistant starch and fibre, but low in fat and animal protein, show a low incidence of colorectal cancer (CRC) of around 1 in 100,000.The taurine to glycine ratio in their conjugated bile acids is around 1:9. By contrast, African Americans consuming Western-style diets high in animal protein and fats show a much higher CRC incidence of around 65 in 100,000 [12]. The taurine:glycine ratios in their conjugated bile acids is around 10:1. This is not to suggest that microbial metabolism of bile acids is solely responsible for the marked difference in CRC incidence. Other factors such as the difference in fibre intake, supply of short chain fatty acids (such as butyrate) and phytochemicals are also relevant, as mentioned already. Nevertheless, microbial metabolism of conjugated bile acids provides a very likely and potentially important mechanism for the promotion of colorectal cancer by diets high in animal fat and protein.

Trimethylamine

Choline (not to be confused with cholic acid, or cholate) is an essential precursor of membrane lipids. In the large intestine, bacterial fermentation of excess choline gives rise to trimethylamine (TMA), which finds its way into the bloodstream and is converted to the compound trimethylamine oxide (TMAO) by the liver. Bacterial formation of TMA from choline has attracted interest because there is an association between TMAO and cardiovascular disease [13]. It is not yet clear however whether TMAO directly causes the disease, or is simply an indicator of it.

Products of Protein Fermentation

Dietary protein is largely broken down to its constituent amino acids and peptides (fragments mostly consisting of two or three amino acids) by our own digestive enzymes such as pepsin and trypsin in the stomach and small intestine. These are then absorbed by in the small intestine and used subsequently to manufacture new proteins in human tissues. Some types of protein are broken down slowly, however, and high dietary protein intakes may lead to incomplete digestion of protein. This means that some dietary protein reaches the large intestine, where it gets fermented by the resident gut microbiota. The gut microbiota also process a wide variety of proteins that are produced by the body as secretions, including digestive enzymes, mucin and antibodies, or derived from cells released by turnover of the gut wall.

The 20 amino acids that make up proteins represent a wide range of chemical structures. Consequently, they give rise to a much wider variety of fermentation products than is the case for sugar fermentation. These include amines that retain the nitrogen-containing amino group, some of which are toxic at high concentrations. On the other hand, removal of the amino group results in ammonia, which is also toxic at high concentrations [14]. The two sulfur-containing amino acids give rise to toxic hydrogen sulfide, while the three aromatic amino acids give rise to aromatic acids that include indoles, phenols and cresol. The branched chain amino acids isoleucine, leucine and valine are fermented to the branched chain fatty acids that were mentioned earlier, and these provide useful indicators of protein fermentation.

The net effect of these products from protein fermentation is largely negative, particularly for gut health, and they may contribute to the increased risk of colorectal cancer that is associated with high protein intakes [15] (Fig. 9.3). Having said this, much of the increased risk that is associated with consumption of red and processed meat may be due to the cancer-promoting (*carcinogenic*) properties of N-nitroso-compounds and haem that are largely independent of microbial activity [16].

Microbial Metabolites and the Brain

Certain metabolites produced by the gut microbiota can directly influence our nervous system and so, potentially, our behaviour. The role of d-lactate as a *neurotoxin* is one example that was mentioned already. Other neuroactive microbial metabolites include compounds such as p-cresol and indole-pyruvate that are derived from the fermentation of aromatic amino acids in the gut. It has been suspected for some time that autism in humans might be associated with changes in gut microbiota [17]. This idea has been given more credence by studies in which symptoms associated with autism (ASD) are induced in mice through a procedure known as 'maternal immune activation' (MIA). The MIA offspring show alterations in their gut microbiome and metabolome, including a particularly large increase in one neuroactive phenolic compound (4EPS). Dosing the MIA offspring with a strain

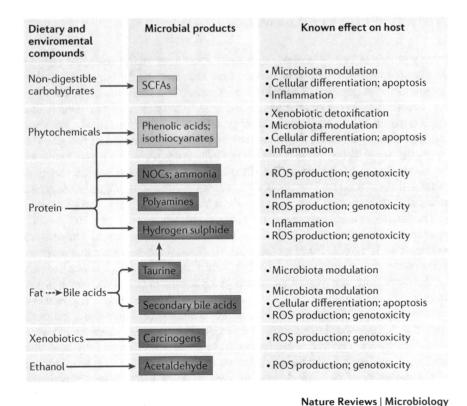

Fig. 9.3 The gut microbiota, bacterial metabolites and colorectal cancer. Metabolites considered beneficial are shown in blue, those considered harmful in orange (reproduced from Louis et al. 2014 [15])

of *Bacteroides fragilis* led to altered microbiota profiles, decreases in circulating metabolites including 4EPS and indole-pyruvate, and alleviation of symptoms associated with ASD [18]. Clearly the changes involved in this mouse model are extremely wide-ranging, involving the immune system, the whole gut microbiome and the serum metabolome. The demonstration that 4EPS can induce the symptoms of ASD in mice, together with the observation that it is elevated along with other neuroactive microbial metabolites in human ASD, however makes microbiota-derived metabolites a plausible mechanism in autism [18].

More generally, since certain microbial metabolites can affect the nervous system, there is broad potential for them to affect our mood, behaviour and cognitive function. Neurotransmitters such as serotonin and GABA (gamma-amino butyric acid), for example, are derived from the aromatic amino acid tryptophan. The production of neurologically active metabolites by the microbiota depends mainly on two things. First, the amounts of protein, fat, carbohydrate and fibre in the food that we eat, which determine the chemical composition of the digesta that reach the microbial community in the large intestine. Second, on the species composition of our own individual gut microbiota, which as we have seen can change with diet, medication and illness. This is an expanding and promising (also challenging) area for research.

Microbial 'Secondary' Metabolites

Gut microorganisms have an extraordinary capacity for chemical conversion and synthesis. This goes well beyond the basic 'housekeeping' functions of gaining energy via fermentation or respiration and synthesising essential cellular components. A range of other products may be formed that are often loosely called *secondary metabolites*. Some of these are small peptides that are produced not through the normal mechanism of protein synthesis, but via *non-ribosomal* synthesis involving special enzyme systems. Gene clusters responsible for producing these and other complex products such as polyketides can be identified in the genome sequences of many gut bacteria [19]. While the roles of these compounds are still being uncovered in detail, they can include antibiotic activity that is directed at other gut microorganisms. Clearly this may have important consequences for competition between microorganisms, including the suppression of pathogens. Other secondary metabolites are involved in various forms of signalling between microbial cells, for example in response to culture density (known as *quorum sensing*). There is also evidence that some microbial secondary metabolites have impacts on host physiology.

Conclusion: The Right Balance

It should be evident that the complex mix of metabolites produced by our gut microbiota contains both protective and potentially damaging compounds. Most of us would say that we are keen to tailor our diets to maximise our chance of living a long life, free of serious illness. While no dietary strategy can guarantee this, the main recommendations currently are to increase fibre consumption and to limit consumption of sugars and of red and processed meat and of fat. Not only are these suggested by the evidence from epidemiology, they are entirely consistent with our understanding of the impact of diet on different aspects of gut microbial metabolism, as discussed in this chapter.

We are now much more aware that variations in our gut microbiota can result in differences between people in their metabolome. As we have seen, these may affect disease risk, responses to food and to drugs, and even behaviour. It is not too fanciful to suppose that we may in future be able to predict and help to avoid some of the negative consequences through profiling of our individual gut microbiota and metabolomes. Indeed, the necessary profiling techniques are already here; rather, it is the connections between individual microbes and metabolites that need to be revealed in greater detail.

References

1. Nicholson JK et al (2012) Host-gut microbiota metabolic interactions. Science 336:1262–1267
2. Cummings JH (1995) Short chain fatty acids. In: Gibson GR, Macfarlane GT (eds) Human colonic bacteria. CRC, Boca Raton, FL, pp 101–130
3. Louis P, Flint HJ (2017) Formation of propionate and butyrate by the human colonic microbiota. Environ Microbiol 19:29–41
4. Reichardt N et al (2018) Specific substrate-driven changes in human faecal microbiota composition contrast with functional redundancy in short chain fatty acid production. ISME J 12:610–622
5. Den Besten G et al (2013) The role of short chain fatty acids in the interplay between diet, gut microbiota and host energy metabolism. J Lipid Res 54:2325–2340
6. Byndloss MX et al (2017) Microbiota-activated PPAR-gamma signaling inhibits dysbiotic Enterobacteriaceae expansion. Science 357:570–575
7. Miyamoto J et al (2016) Nutritional signalling via free fatty acid receptors. Int J Mol Sci 17:45
8. Chambers ES et al (2015) Effects of targeted delivery of propionate to the human colon on appetite regulation, body weight maintenance and adiposity in overweight adults. Gut 64:1744–1754
9. Braune A, Blaut M (2016) Bacterial species involved in the conversion of dietary flavonoids in the human gut. Gut Microbes 7:216–234
10. Duncan SH et al (2016) Wheat bran promotes enrichment within the human colonic microbiota of butyrate-producing bacteria that release ferulic acid. Environ Microbiol 18:2214–2225
11. Ridlon JM, Wolf PG, Gaskins HR (2016) Taurocholic acid metabolism by gut microbes and colon cancer. Gut Microbes 7:201–215
12. O'Keefe SJ et al (2015) Fat, fibre and cancer risk in African Americans and rural Africans. Nat Commun 6:6342
13. Schiattarella GG et al (2017) Gut microbe-generated metabolite trimethylamine-N-oxide as cardiovascular risk biomarker: a systematic review and dose-response analysis. Eur Heart J 38:2948
14. Windey K et al (2012) Relevance of protein fermentation of gut health. Mol Nutr Food Res 56:184–196
15. Louis P et al (2014) The gut microbiota, bacterial metabolites and colorectal cancer. Nat Rev Microbiol 12:661–672
16. World Cancer Research Fund/American Institute of Cancer Research (2007) Food, nutrition, physical activity and the prevention of cancer: a global perspective. http://wcrf.org/int/research-we-fund/continuous-update-project-cup/second-expert-report
17. Finegold SM et al (2010) Pyrosequencing study of association of fecal microbiota of autistic and control children. Anaerobe 16:444–453
18. Hsiao EY et al (2013) Microbiota modulate behavioural and physiological abnormalities associated with neurological disorders. Cell 155:1451–1463
19. Donia MS et al (2014) A systematic analysis of biosynthetic gene clusters in the human microbiome reveals a common family of antibiotics. Cell 158:1402–1414

Chapter 10
Host Responses to Gut Microbes

Host Defence Systems

So far, we have tended to emphasise the benefits of symbiotic association between the host and its gut microbiota. On the other hand, we know only too well that micro-organisms and viruses pose a continual threat to the survival of multicellular life forms, and these micro-organisms include many that either inhabit, or pass through, our gut. A wide range of viruses, bacteria and fungi are pathogens that have the ability to infect particular host species, while many protozoal species are pathogenic or parasitic. This has imposed incredibly strong selection for systems and mechanisms that help the host animal to defend itself against attack from micro-organisms and viruses.

A level of protection is provided against food-borne pathogens by the acidic stomach and against gut micro-organisms in general by the flow of gut contents (peristalsis), by the mucus layer and by production of anti-microbial peptides known as 'defensins'. The ultimate defensive system however is the immune system. In humans and other mammals there are two distinct systems. Versions of the *innate immune system* are also found in invertebrate animals, including insects and the coelenterate *Hydra*, indicating an early evolutionary origin. At the core of this system are "pattern recognition receptors", which are proteins on the surface of many cell types that recognize molecules or structures associated with microbial cells. In mammals, particular receptors known as TLRs (toll-like receptors) recognize bacterial flagella, lipoteichoic acid found in Gram-positive bacterial cell walls, lipopolysaccharides found in Gram-negative bacterial cell walls, and even particular nucleotide sequences (containing unmethylated CpG) that occur in bacterial DNA. These provide early warning of the presence of potentially infectious micro-organisms and viruses and trigger an initial set of responses to resist and combat infection. These responses include recruitment of white blood cells (called 'phagocytes') that are able to engulf infectious microbes and activation of a system called the 'complement cascade' that helps to destroy microbial cells.

© Springer Nature Switzerland AG 2020
H. J. Flint, *Why Gut Microbes Matter*, Fascinating Life Sciences,
https://doi.org/10.1007/978-3-030-43246-1_10

Fig. 10.1 Structure of an
Immunoglobulin molecule.
Each molecule is composed
of two long (heavy) protein
chains and two short (light)
protein chains as shown.
The amino acid sequences in
the 'Y' shaped region that
recognizes antigens are
highly variable

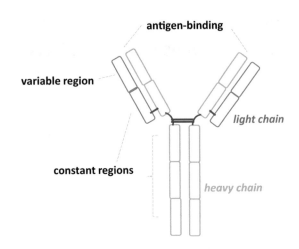

    Luckily, we also possess a second, highly sophisticated system known as *adaptive immunity*. This system has the remarkable ability to learn to recognise and respond to microbial challenges in a very specific, targeted, manner. Not only this, the adaptive immune system has the ability to remember past infections, so that it is able to prevent them altogether when the same infectious organism or virus is encountered again at a later time. This is the whole basis for our highly effective vaccination programmes, in which the body is deliberately exposed to distinctive cell structures (known as 'antigens') from particular micro-organisms or viruses, delivered in a harmless way. Most commonly, antigens are fragments of proteins or polysaccharides. In this way we can take full advantage of our immune system by ensuring that it gets to recognise the infectious agent and so prevent future infection, without the body ever having to experience the infection itself.

    The primary weapons of the adaptive immune system are the Y-shaped immunoglobulin proteins that we call antibodies (Fig. 10.1). The main producers of antibodies that act against free viruses and bacteria are white blood cells called B lymphocytes. These are initially produced in the bone marrow and then become activated to produce antibodies by exposure to antigens in the spleen and lymph nodes. A second class of T lymphocytes are produced in the thymus and mainly target human cells that are expressing foreign antigens because of viral infection or cancer.

    By binding to potentially infectious microbial cells and viruses, antibodies help to trigger a series of events that lead to their destruction. Being proteins, antibodies are coded for by genes present in our chromosomes and are manufactured on ribosomes. The ability of an antibody to recognise (i.e. bind to) a particular microbial or viral antigen depends on the precise sequence of amino acids that is present on the arms of the immunoglobulin molecule. Because twenty different amino acids can be involved in a protein chain, and these amino acids can occur in almost any order, the potential number of different protein sequences is truly astronomical. This creates the possibility for a vast number of different antibody sequences to exist, each of which may recognise a different antigen.

But there is an obvious problem here. How can our chromosomes possibly carry enough genes to produce immunoglobulin proteins that specifically recognise all of the different infectious micro-organisms and viruses that we may be exposed to during our lifetimes? The answer is that they simply could not, and indeed they do not. What actually happens is that during the development of antibody-producing cells (the lymphocytes), parts of their immunoglobulin genes get shuffled (recombined) rapidly so as to create a range of alternative sequences. In this way, a diverse set of immune cells is created that produce antibodies with different binding specificities. The trick then is for the system to select those cells with the right binding specificity for the 'job-in-hand', i.e. coping with the threatened infection. This is exactly what happens when we are exposed to an initial infection, or to a harmless vaccine. The consequence is that only the particular 'clone' of cells with the appropriate binding specificity is amplified.

An antibody response to a new antigen happens in two phases. Upon first exposure to the antigen there is a primary response that takes up to a week to reach peak antibody levels. It is during this time that the proliferation of antibody-producing lymphocytes occurs, along with selection of those cells that have the appropriate immunoglobulin sequences to bind the antigen. Importantly, some of these lymphocytes persist as "memory cells" after the infection and primary response have passed. It is these cells that provide the starting point for the secondary immune response. This means that when the secondary response is triggered by a subsequent exposure, it follows the antigen challenge more quickly and produces far more antibody (by 100–1000 times) over a longer period of time than was the case in the primary response. Furthermore, these antibodies are now tailored and targeted to the particular infection. The predominant immunoglobulin produced in the secondary response is of a different type (IgG) to that mainly involved in the primary response (IgM).

Another obvious question is, given that our own cells and tissues are made up of huge numbers of molecules and structures that represent potential antigens, why do these not themselves provoke an immune response? The answer is that the immune system has somehow to learn to recognise these structures as 'self' and so avoid reacting to them. This process is known as 'immune tolerance' and is obviously of enormous importance in preventing inappropriate antibody responses. 'Central tolerance' involves the editing out during their development of lymphocyte clones that recognise any antigens they encounter in their tissue of origin (bone marrow, thymus). This is therefore a key mechanism for preventing responses to our own antigens. It is also supplemented by 'peripheral tolerance', which acts on mature B and T lymphocytes and involves a special group of T lymphocytes known as regulatory T cells (Tregs). Tregs can be thought of as a sort of immunological police force that acts to control inappropriate immune responses. Unfortunately, things can go badly wrong with immune tolerance mechanisms. When this happens, it leads to what we know as autoimmune diseases (such as rheumatoid arthritis, lupus and others) in which the immune system attacks the wrong targets.

Immune Responses to Gut Microbiota

There are roughly as many microbial cells in our gut as there are human cells in the whole body. Thus, the gut microbiota represents a vast pool of potential antigens all of which belong to 'foreign' cells. In general, however, the immune system does not waste its resources in reacting to harmless micro-organisms in the gut. As we will see, there are many reasons for this. The first reason is that the healthy gut lining is a very effective barrier that prevents access of gut micro-organisms to body fluids and tissues. This means that under normal circumstances only a few specialised pathogens that have the ability to invade host tissues or seriously damage the gut lining can reach the bloodstream and elicit an immune response. When the gut barrier is compromised, however, even normally harmless resident gut microorganisms can enter the bloodstream and provoke an immune response.

On the other hand, the immune system is certainly not 'blind' to antigens present in the gut. Particularly in the small intestine there are regions of the gut mucosal surface that contain special 'M' cells which transport antigens from the gut contents so that they are presented to immune cells within the extensive gut-associated lymphoid tissue (GALT). Dendritic cells (so-called because they have projections, or 'dendrites') are also involved in this sampling and presentation of microbial antigens from the gut. The outcome is a constant surveillance of the gut microbiota that can lead either to tolerance of the resident micro-organisms, or to early warning of potential infection and the need for an immune response. Some 3–5 g of Immunoglobulin A (IgA) antibody is secreted into the intestine per day, where it can be seen to coat bacterial cells [1]. This secreted IgA helps to neutralise bacterial toxins and to prevent harmful bacteria from coming into contact with the gut mucosa. It is thought there may also be a role in promoting tolerance of commensal bacteria [2].

It is also important to note that the development of the immune system through early life depends on its exposure to microbial antigens in the gut. We know this because 'germ-free' rodents show abnormal development of their immune systems and even of their gut anatomy [3]. This is the basis for what has been called the 'hygiene hypothesis' that attributes many allergies and autoimmune conditions to the success of modern approaches to hygiene, resulting in a lack of early exposure to microbial antigens [4]. It should be stressed however that this proposal is still being debated, and that improved hygiene has greatly reduced early mortality resulting from infectious disease.

Human Genetic Variation

All the elements of the highly complex immune system are coded for by our genes. But gene copies can exhibit polymorphism, meaning that they often vary in their precise sequences between individuals. Probably the most familiar examples of

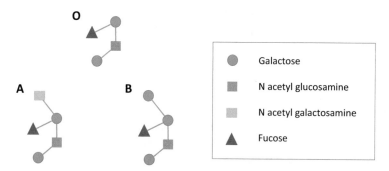

Fig. 10.2 Human ABO blood group polymorphism. Three different antigens are determined by three different arrangements of sugars on the surface of cells, governed by our genes (genotype). Those with an *OO* genotype are blood group O, *AA* or *AO* genotype blood group A, *BB* or *BO* genotype blood group B, and *AB* genotype blood group AB

genetic polymorphism are the ABO blood groups (Fig. 10.2). The products of these genes determine which carbohydrate groups are present on cell surfaces and it is these carbohydrate antigens that cause incompatibility between donor and recipient blood. Tissue incompatibility is determined by highly polymorphic genes of the major histocompatibility complex (MHC), which code for proteins that are important in immune recognition. Not surprisingly, polymorphism can also have major consequences for the relationship between our gut microbiota and the immune system. For example, it is well established that susceptibility to Crohn's disease, which is one type of inflammatory bowel disease, is partially inherited [5]. Many different genes are involved, but one of those with major effects is the *nod2* gene that codes for a component of the innate immune system. In individuals who have two defective copies of *nod2*, the normal recognition of a particular bacterial cell surface structure and the consequent upregulation of antimicrobial defensin peptides produced at the mucosal surface, are missing [6]. This contributes to imbalance in the species composition of the gut microbiota and to the severe inflammation that characterises Crohn's disease.

Trade-offs Between Tolerance and Responsiveness

Combatting harmful micro-organisms comes at a considerable cost to the body in terms of energy expenditure (not least to synthesise all the extra immune cells and proteins involved), tissue damage, inflammation, body temperature and pain. Clearly in the case of life-threatening infections a strong immune response is essential. For some more minor chronic conditions, however, tolerance might offer a better alternative strategy, and this may indeed be what has led on to co-existence and commensalism. The balance of this trade-off has been determined through the evolution of immune responses and of immune tolerance mechanisms. An important factor must be that an established and co-adapted commensal microbiota itself provides significant protection against pathogenic invaders. This happens through competition for nutrients, control of the gut environment (especially pH) and the production of growth-inhibitory compounds by commensal bacteria. By tolerating these commensal organisms, the host therefore gains an important additional defence mechanism without a heavy cost to itself. The close interdependence of the immune system and the resident gut microbiota is demonstrated by the fact that mice bred and reared under germ-free conditions show abnormal development of their gut lining, with minimal lymphoid tissue and antibody secretion.

Given that bacteria replicate perhaps 20,000 times faster than we do (their generation times being measured in hours rather than years), they will always have the capacity to evolve far more rapidly than us. This has allowed them to subvert many of the body's defence mechanisms. For example, some pathogenic bacteria actually use the M cells of the small intestine, which exist largely to monitor intestinal antigens, as their route of entry for infection. Other pathogens have the ability to switch the major antigens presented on their cell surface, as in *Bacteroides fragilis* where this switching involves a genetic shuffling mechanism [7]. Meanwhile, viruses mutate so rapidly that new vaccines have to be prepared against the most recent strains of human influenza to emerge every year. The challenges posed by increasing human population density and travel volumes mean that the threat from new pathogenic micro-organisms and viruses is unlikely to decrease in the future.

To What Extent Does Our Own Genome Determine Our Gut Microbiome?

It is clear that human genetic variation involving components of the immune system can affect the composition of our gut microbiome. But this is just one of many ways in which variation in our genomes may influence the micro-organisms that colonise our intestine. For example, the amounts and types of digestive enzymes that we produce (including amylases, lactases, proteases) must influence what digestive residues reach the large intestine. Also, the precise structures of carbohydrates on the surface of cells lining the intestine, or present in human milk oligosaccharides (HMOs), will determine whether certain bacteria can bind to them, or gain energy by degrading them. Thinking more broadly, the anatomy and physiology of our digestive system, which governs transit, peristalsis and the regulation of gut pH are determined by our genes. The same is largely true for our eating behaviour which determines, among other things, how much fibre we like to consume.

The revolution in DNA sequencing methods (Chap. 3) now makes it possible to detect human genetic variation across an individual's whole genome. Since we can also describe that individual's microbiome in molecular detail, we can ask whether there is any link between the two. If this is done for large enough groups of people, we can discover whether certain human gene variants (polymorphisms) tend to be preferentially associated with particular groups of micro-organisms within the microbiome. Crucially, there is also the potential to link both the polymorphisms and microbiome composition to traits that are related to human health. We mentioned lactose intolerance in Chap. 7 and can add here that the genetic basis for lactose intolerance is largely determined by the genetic variation in the human *LCT* gene which codes for the enzyme lactase. It turns out that the *LCT* gene variants are associated not only with dairy intake, but also with the incidence of bifidobacteria within the gut microbiota [8]. This conclusion is also supported by studies with monozygotic twins, which of course have identical genomes [9]. Genome-wide association studies of this type have indicated that at least a third of bacterial taxa within our gut microbiome show significant correlations with host genotype [10]. These include such important and abundant groups as *Faecalibacterium* and *Bifidobacterium* species. We can expect many exciting and important developments through such approaches in the coming years.

Conclusions

This chapter has presented a very rapid overview of the highly complex interaction between our immune system and our gut microbiota. This interaction has been shaped by powerful selective forces operating throughout human history, imposed mainly by the continual challenge of microbial, viral and parasitic infections. It is obvious that lethal epidemics tend to remove genes from the population that confer sensitivity and select for genes that promote survival against the infectious agent. The genes selected for by such extreme events may however turn out to have negative consequences in the absence of infection. The classic example of this is sickle cell anaemia, where the possession of one 'sickle cell' gene affords some protection again malaria even though two copies of the sickle cell gene cause anaemia [11]. This is a case of what is known as 'balancing selection' in which alternative versions of a gene are promoted at different times, depending on the circumstances and the *heterozygous* state is more advantageous for survival than either *homozygous* state. This phenomenon is considered to be a major cause of genetic polymorphism in the human population, including the genes that specify components of the immune system.

References

1. McPherson AJ et al (2008) The immunogeography of IgA induction and function. Mucosal Immunol 1:11–22
2. Corthesy B (2013) Role of secretory IgA and maintenance of homeostasis. Autoimmun Rev 12:661–665
3. Imhann F et al (2018) Interplay of host genetics and gut microbiota underlying the onset and clinical presentation of inflammatory bowel disease. Gut 67:108–119

4. Olszak T et al (2012) Microbial exposure during early life has persistent effects on natural killer T cell function. Science 336:489–493

5. Rautova S et al (2004) The hygiene hypothesis of atopic disease – an extended version. J Pediatr Gastroenterol Nutr 38:378–388

6. Sidiq T et al (2016) Nod2: a critical regulator of ileal microbiota and Crohn's disease. Front Immunol 7:367

7. Cerdena-Tarraga AM et al (2005) Extensive DNA inversions in the B. fragilis genome control variable gene expression. Science 307:1463–1465

8. Bonder MJ et al (2016) The effect of host genetics in the gut microbiota. Nat Genet 48:1407–1412

9. Goodrich JK et al (2016) Genetic determinants of the gut microbiome in UK twins. Cell Host Microbe 19:231–243

10. Turpin W et al (2016) Association of host genome with intestinal microbial composition in a large healthy cohort. Nat Genet 48:1413–1417

11. Allison AC (1956) The sickle-cell and Haemaglobin C genes in some African populations. Ann Hum Genet 21:67–69

Chapter 11
Treating the Gut Microbiome as a System

'Molecular Biology' has dominated the study of biology and microbiology since publication of the structure of DNA in the 1950s [1]. Right from the start, Molecular Biology, which focusses on the flow of genetic information from DNA to RNA to protein, produced staggering conceptual and technical advances. These included unravelling the genetic code by which nucleotide sequences are translated into protein sequences and the mechanisms that control when and where genes are expressed. On the technical side, the 'recombinant DNA revolution' was created by the ability to artificially stitch together (recombine) stretches of DNA of different origin (e.g. bacterial and mammalian) and then to propagate and manipulate them almost at will. Furthermore, in recent years the ability to sequence DNA molecules has increased in speed and decreased in cost at an astonishing pace.

In science, as in other areas of human activity, funding and expertise tends to flood towards those topics that are the fastest moving. Initially this led to focus being on a small number of 'model' organisms, notably the bacterium *Escherichia coli* and the baker's (and brewer's) yeast *Saccharomyces cerevisiae*. These organisms were originally studied because of their medical and economic importance, respectively, but also because, as oxygen-tolerant species, they are easy to grow and maintain in culture. The biochemistry, genetics and cell biology of these organisms have been pursued exhaustively, with excellent results. Recombinant DNA technology developed largely as a spin-off from fundamental work with *E. coli*, with '*vectors*' used to clone DNA fragments being derived from plasmids and bacteriophage viruses associated with *E. coli*. Likewise, the enzymes that are used to cut, re-join and replicate DNA molecules in the test-tube ('restriction enzymes', DNA ligase, DNA polymerase) were originally identified from work in *E. coli*. Only relatively recently has scientific focus shifted towards the huge number of less fashionable micro-organisms, and towards complex microbial communities, with this shift being driven largely by the stellar advances in DNA sequencing technology.

An inherent danger of this tendency towards 'herd behaviour' among scientists is that important but unfashionable (or un-fundable) approaches and ways of thinking can get lost. The huge success and relative ease of sequence-based approaches can

© Springer Nature Switzerland AG 2020
H. J. Flint, *Why Gut Microbes Matter*, Fascinating Life Sciences,
https://doi.org/10.1007/978-3-030-43246-1_11

lead to a somewhat one-dimensional approach in which the only meaningful answers to biological questions are assumed to derive from determining and then analyzing nucleotide sequences. An alternative to this wholly reductionist approach is offered by treating complex biological networks instead as systems. 'Systems Biology' has now become a recognized discipline, although it means rather different things to different people. We will first consider the original meaning of Systems Biology before touching on some of the alternative interpretations that have emerged more recently.

One of the first proponents of the approach that has come to be known as Systems Biology was my own PhD supervisor, Henrik Kacser [2]. One phenomenon that is neatly explained by his systems approach is *genetic dominance*, first identified by the father of genetics, Gregor Mendel, in the 1860s. Despite Mendel's work, it had remained a complete puzzle why in diploid organisms like ourselves (which carry two copies of each chromosome and therefore of each gene), possession of a normal gene copy generally cancels out (i.e. dominates) the effects of a second, defective gene copy. Since most genes code for enzyme proteins, another way to state this is that, in most cases, we can function perfectly well with only half the normal activity of a given enzyme (assuming that the defective gene produces little or no active enzyme). But why should this be? Well, most enzymes contribute to lengthy biochemical pathways in which multiple enzymes combine to make a particular product (e.g. an amino acid or a nucleotide base). Elegant mathematical analysis proved that control of the flow (or flux) through the pathway must be shared between the various pathway enzymes, such that decreases of up to 50% in the activity of the 'average' enzyme in the pathway would have only a minimal effect on the formation of the product [3]. This is largely because of the biochemical 'buffering' that is inherent in such pathways. As can be shown experimentally, a decrease in one pathway enzyme results in an increased concentration of its substrate, which itself causes a compensatory increase in the rate of the reaction [4] (Fig. 11.1). In other words, the phenomenon of genetic dominance turns out to be a straightforward prediction arising from a proper quantitative approach to biochemistry [5]. The key point here is that dominance is a property of the **system** rather than the individual gene or enzyme, as it depends on the biochemical context in which the enzyme operates.

Although it is something of a leap to go from biochemical pathways in a single organism to the consideration of microbial communities, the basic principles of the systems approach are equally relevant at the community level. As we have seen already, the metabolic products and combined activities of the whole microbial community in the gut have a major impact on the nutrition and health of the host organism. Very few of these products arise from the activity of a single organism. More often they result from the combined activities of many members of the community. To understand this, we need some basic information on the individual micro-organisms that make up the community, or rather on groups of micro-organisms (such as methanogenic archaea, or butyrate-producing bacteria). This is most reliably obtained from cultured representatives. Thereafter, we need to try to predict how functionally different groups will interact with each other and what

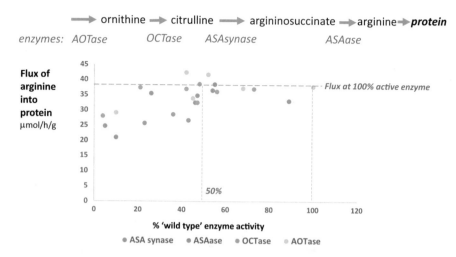

Fig. 11.1 How far can you decrease an enzyme activity before it affects the final output of a pathway? The activity of enzymes responsible for making the amino acid arginine (which is essential for protein synthesis) was varied experimentally in the mould *Neurospora crassa*. Decreases in four successive pathway enzymes of up to 50% are shown here to have little effect on the flux of arginine into protein (Flint et al. 1981 [4]). This behaviour is predicted by the summation theorem of Kacser and Burns 1973 [3] and underlies the phenomenon of genetic dominance (Kacser and Burns 1981 [5])

products will be formed by the microbial community under different conditions of pH, turnover and substrate supply. Initial approaches to these questions have now been made by theoretical modelling as will be discussed below.

Modelling of Gut Microbial Communities

Experimental microbial communities can be set up using an approach known as the *chemostat*. In its simplest form this consists of a fixed volume of growth medium within a chamber. Fresh sterile medium is fed in, and the same volume of used medium removed, at a constant rate known as the *dilution rate*. Bacteria are inoculated into the chamber (either as a pure culture or as a mixture) and the nutrients in the medium support their growth. Bacterial numbers will increase initially but then gradually stabilise. When the increase in cell numbers exactly matches the removal of cells by dilution, the culture is said to be in a '*steady state*'. The rate of cell growth is determined by the steady state concentration of the most growth-limiting nutrient in the medium within the chamber, which is usually the main energy source (typically a carbohydrate). Chemostats provide a very valuable tool for investigating the behaviour of microbial communities. They can be run under anaerobic conditions and inoculated with mixed faecal bacteria so as to study the human colonic microbiota as a whole. This has allowed us to study the response of

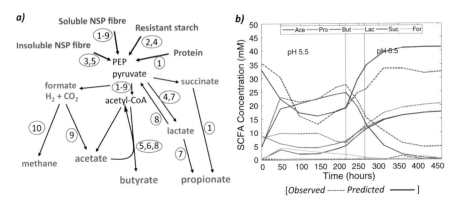

Fig. 11.2 Theoretical modelling of the human colonic microbiota. (**a**) Modelling of short chain fatty acid formation by subdividing the microbiota into ten functional groups (1–10) with specified substrate preferences and products (substrates in red, products in blue); (**b**) application of this approach to model the impact of a one unit pH shift upon microbial metabolites in a chemostat community (Walker et al. 2005 [6]; Kettle et al. 2015 [8])

human gut communities to shifts in pH and the type of carbohydrate substrate supplied under carefully controlled conditions [6, 7].

An advantage of this experimental approach is that 'steady state' data are very amenable to theoretical modelling. By avoiding the complications that arise from changes with time and location, the theoretical analysis is simplified and it is possible to focus on the microbial community and its metabolic products. We have developed a theoretical model of the human gut microbiota to simulate some actual chemostat experiments [8]. To achieve this, our 'virtual' gut microbiome was subdivided into ten functional groups (microbial functional groups, or MFGs). The MFGs were assumed to differ in the substrates that they utilise and in the fermentation products that they produce, based on existing knowledge mainly from cultured isolates (as discussed in Chaps. 4, 5 and 7) (Fig. 11.2a). These attributes are defined in the model by specifying the relevant biochemical pathways and by assigning maximum growth rates and substrate affinities for each MFG. The system is described by a set of differential equations, and predictions about its behaviour are made through computer modelling (the model is called 'microPop') [9].

Our first application of this approach was aimed at understanding the potential effects of pH change upon the human gut microbiome. Our chemostat experiments (in which the pH is under experimental control) had shown that a slightly acidic pH can favour the growth of butyrate-producing Firmicutes within the community. On the other hand, switching to a near neutral pH led to greater dominance of *Bacteroides* species, a higher proportion of propionate and less butyrate [6]. In our theoretical modelling, we introduced a pH response factor to describe the response of different MFGs to changing pH, with *Bacteroides* species showing more sensitivity than butyrate-producing Firmicutes to growth inhibition by a slightly acidic pH (as is observed for representative pure cultures [10]). In the event, our model predictions came remarkably close to the experimental observations on community composition and short chain fatty acid formation [8] (Fig. 11.2b).

Our model specifies the exact relationships between growth substrates and products of fermentation (known as 'reaction stoichiometries') for each MFG. In some cases, these relationships are not fixed and we find that pure cultures of many butyrate-producing bacteria form more butyrate at a slightly acidic pH [11]. This also applies to the whole community, as we find that more butyrate is produced per butyrate-producing bacterium in mixed incubations at pH 5.5 than at pH 6.5 [12]. Thus, both increased populations of butyrate-producing bacteria and shifts in their biochemistry contribute to increasing butyrate production at the lower pH. Both aspects can be accounted for in the model. It has been known for some time from human volunteer studies involving consumption of different fibres, that higher faecal SCFA concentrations are accompanied by a disproportionate increase in butyrate [13]. Likewise, we find that decreased total SCFA concentrations resulting from lower fibre intakes are accompanied by a disproportionate decrease in faecal butyrate [14]. These changes are almost certainly explained by the decrease in pH that results from the fermentation of fibre to SCFA in the proximal colon. The modelling has helped us to understand exactly why it is that lower pH promotes butyrate formation by the microbial community.

Theoretical modelling of the microbial community achieves several things [15]. Most of all, it allows us to rigorously test the validity of our hypotheses and assumptions. With highly complex systems, one has to be wary of making simplistic 'off-the-cuff' predictions about the outcome of manipulations and challenges, as these can often turn out to be counter-intuitive. The power and precision of computer modelling should therefore be indispensible in making predictions. Inevitably some model predictions will not match the experimental observations, which may indicate missing information or assumptions that need revising. A very important function of the model then is to define areas of ignorance and so inform future research. Modelling is therefore an iterative process as new experimental findings should lead constantly to modifications in the basic assumptions made. A good example is that we now recognise, based on evidence discussed in Chap. 6, the urgent need for an additional MFG in order to distinguish the behaviour of *Prevotella* from that of *Bacteroides* species.

Further refinements may be needed to build the role of '*keystone species*' into our models. Keystone species are ones whose absence has a wide-ranging impact on the whole community. As noted in Chap. 7, only a few species within the community may be capable of initiating the breakdown of insoluble fibres such as lignocellulose or resistant starch [16]. In their absence, this material remains largely undegraded, thus eliminating a major flow of carbon and energy to the rest of the community. Another example of keystone species are the specialist anaerobes that remove lactic acid by converting it to butyrate or propionate (Chap. 4). Because accumulation of lactic acid lowers gut pH and can destabilise the whole community, the system is highly sensitive to any change in its rate of removal. While the populations of such keystone species may be numerically small, they will need to be recognized individually as separate MFGs.

Modelling the Gut Itself

A chemostat community does not, of course, provide a model of the gut itself. There are a great many reasons for saying this. Most obviously, the lack of host cells means that interactions between the microbial community and the gut wall, the immune system and host secretions (including mucin) are either absent or very hard to reproduce. For example, we know that the gut wall is responsible for constantly absorbing short chain fatty acids and water from the gut contents. This process cannot easily be simulated using in vitro (literally 'in glass') systems, although heroic attempts have been made to remove SCFA in some more complex systems by dialysis [17]. One consequence of the lack of SCFA removal is that microbial cell densities are never quite as high as can be attained in the large intestine, since high SCFA concentrations themselves tend to limit microbial growth. Also, the lumen of the gut does not consist of a single 'well-stirred' space. Rather the gut is a lengthy tube whose diameter, throughput, environment (pH) and microbiota differ in different regions. Again, heroic attempts have been made to construct complex in vitro systems consisting of multiple compartments arranged in series with the aim of simulating digesta passage through different regions of the gut. We know that the contents of the gut can never approach a true 'steady state' as is approximated in a chemostat. In part this is because food intake (at least in humans) is periodic. Most of us eat meals only at certain times of the day, so that the supply of digesta passing through the gut towards the large intestine will also be periodic. On the other hand, much of our data, particularly for healthy subjects, continues to be derived from the analysis of faecal samples. Because of the long residence times in the large intestine, faecal samples represent a long term 'average' of events that have occurred over time down the length of the gut. It can reasonably be argued that steady-state theoretical treatments still provide a convenient way to approximate the outcome of these events.

Theoreticians have now turned their attention to models that consider SCFA and water absorption, periodic feeding, and the effects of different flow patterns and of peristalsis upon microbial fermentation in the human large intestine [18]. In combination with more sophisticated modelling of the gut microbiota, this should ultimately help to interpret the increasing amounts of data emerging from human studies.

Animal 'models' of course provide in vivo gut systems with real gut anatomy, nutrient absorption, peristalsis and immune function. Rodent models are very widely used but they also suffer from significant limitations. The species composition of the rodent gut microbiota differs significantly from that of the human gut and there are important differences in immune function, digestive anatomy and eating habits (not least of which is the propensity of rodents for coprophagy—or eating their own faeces). This makes it difficult to be sure that what is found in rodents always applies to humans, even when gnotobiotic animals implanted with human colonic bacteria are used. The need for definitive human studies is therefore clear and it is essential that we find the best approaches to interpreting the sometimes challenging data that arise from them.

'-Omics' Based Systems Analysis

The important thing about the type of systems approaches discussed above is that they are truly **quantitative**, in that they aim to predict actual cell populations, metabolite production rates and concentrations based on microbial growth rates, growth yields and substrate preferences. Entirely different approaches that have also been described as 'Systems Biology' have been taken by many groups in deploying the -Omics-based technologies that we have discussed in Chap. 3 and elsewhere. These take advantage of powerful new techniques such as metagenomics and metabolomics to generate vast amounts of data from large numbers of samples (e.g. [19]). The 'systems' element here refers first to the fact that potentially all microorganisms and all metabolites within the system (biological sample) can be measured simultaneously. On top of this, however, increasingly sophisticated data analysis (bioinformatics) allows the complex datasets to be searched extensively for correlations. These may be between different organisms (co-occurrence), or between micro-organisms and metabolites, or between any of these variables and recorded characteristics of the individuals sampled (body mass index, state of health, age, gender etc.). This type of analysis clearly has huge potential value for generating hypotheses about possible causal relationships. On the other hand, these '-Omics based' approaches for the most part provide descriptive 'snapshots' of the complex system. One particularly intriguing development is the ability to deduce microbial growth rates in vivo from metagenome data. This is possible because circular bacterial chromosomes are replicated from a fixed point (the origin). As a result, in growing bacterial cultures, genes closer to the origin will be represented by more copies than those further away—and the faster the growth of the culture, the larger the difference in copy number [20].

Another area of interest is the prediction of biochemical pathways from sequence data. For example, the genomes of isolated human gut bacteria have been used to predict the potential for vitamin synthesis and conversely the likely vitamin requirements for these bacteria [21]. Such analysis can also be applied to genomes that are assembled from uncultured organisms, and so may be of assistance in designing new culture media with which to isolate them. These approaches assume of course that we can identify all the relevant genes and pathways by their similarity to those in the few organisms that have been studied intensively in culture (i.e. that we already know all the biochemical steps and gene-protein relationships that there are to know about!). In practice, our knowledge is still incomplete even for these familiar model organisms and the possibility that new pathways and alternative enzyme proteins exist in less-studied, or uncultured, organisms is very real. Functional studies on cultured organisms and gene products still provide the main route for obtaining fundamentally new information.

Conclusions

We are inundated with increasing volumes of detailed information, much of it sequence-based, on gut communities. At the same, many aspects of health and nutrition depend on metabolite concentrations and flow rates that are not so easy to measure directly, or to predict from sequence data. Systems approaches based on computer-assisted modelling appear essential not only for organizing this deluge of information, but also for trying to predict what it all means for our dietary regimens and our health. Theoretical modelling should prove extremely helpful in interpreting the data that can be obtained most readily on samples of faeces, blood and urine.

References

1. Watson JD, Crick FHC (1953) Molecular structure of nucleic acids – a structure for deoxyribose nucleic acid. Nature 171:737–738
2. Martynoga B (2018) Molecular tinkering: the Edinburgh scientists who changed the face of modern biology. CPI Group (UK) Ltd, Croydon
3. Kacser H, Burns JA (1973) The control of flux. Symp Soc Exp Biol 27:65–104
4. Flint HJ et al (1981) Control of the flux in the arginine pathway of *Neurospora crassa*: modulations of enzyme activity and concentration. Biochem J 200:231–246
5. Kacser H, Burns JA (1981) The molecular basis of dominance. Genetics 97:639–666
6. Walker AW et al (2005) pH and peptide supply can radically alter bacterial populations and short chain fatty acid ratios within human colonic microbial communities. Appl Environ Microbiol 71:3692–3700
7. Chung WCF et al (2016) Modulation of the human gut microbiota by dietary fibres occurs at the species level. BMC Biol 14:3
8. Kettle H et al (2015) Modelling the emergent dynamics and major metabolites of the human colonic microbiota. Environ Microbiol 17:1615–1630
9. Kettle H et al (2018) microPop: mathematical modelling of communities in R. Methods Ecol Evol 9:399–409
10. Duncan SH et al (2009) The role of pH in determining the species composition of the human colonic microbiota. Environ Microbiol 11:2112–2122
11. Louis P, Flint HJ (2017) Formation of propionate and butyrate by the human colonic microbiota. Environ Microbiol 19:29–41
12. Reichardt N et al (2018) Specific substrate-driven changes in human faecal microbiota compostion contrasts with functional redundancy in short chain fatty acid production. ISME J 12:610–622
13. Cummings JH et al (1978) The colonic response to dietary fibre from carrot, cabbage, apple, bran and guar gum. Lancet 1:5
14. Duncan SH et al (2007) Reduced dietary intake of carbohydrates by obese subjects results in decreased concentrations of butyrate and butyrate-producing bacteria in feces. Appl Environ Microbiol 73:1073–1078
15. Widder S et al (2018) Challenges in microbial ecology: building predictive understanding of community function and dynamics. ISME J 10:2557–2568
16. Ze X et al (2013) Some are more equal than others: the role of "keystone" species in the degradation of recalcitrant substrates. Gut Microbes 4:236–240
17. Venema K, van den Abbeele P (2013) Experimental models of the gut microbiome. Best Pract Res Clin Gastroenterol 27:115–126

18. Cremer J et al (2017) Effect of water flow and chemical environment on microbiota growth and composition in the human colon. Proc Natl Acad Sci USA 114:6438–6443
19. Nicholson JK, Lindon JC (2008) Systems biology: metabonomics. Nature 455:1054–1056
20. Korem T et al (2015) Growth dynamics of gut microbiota in health and disease inferred from single metagenomic samples. Science 349:1101–1106
21. Magnusdottir S et al (2015) Systematic genome assessment of B-vitamin biosynthesis suggests co-operation among gut microbes. Front Genet 6:148

Chapter 12
Perspectives and Prospects

We have seen how developments in molecular biology, chemical analysis and bioinformatics provide unprecedented new tools for analysing microbial communities and their activities. We have also seen how these are revealing the impact that microbial colonisation of the gut has on many aspects of human health (Fig. 12.1) [1]. So, perhaps the most obvious question that we should be asking is—exactly what composition of the gut microbiota is most beneficial for health? This is clearly a key question but, despite a deluge of new data, it is surprisingly difficult to provide a simple, straightforward answer.

What Is the 'Ideal' Composition of Our Intestinal Microbiota?

The most obvious characteristic of a 'healthy' gut microbiota must be the absence of harmful agents, which can include infectious viruses, pathogenic bacteria and eukaryotic micro-organisms. So, does this mean that we would all benefit from having a sterile intestine with no micro-organisms at all (however difficult this might be to achieve)? The answer is a resounding 'No'. We have seen that many of the predominant commensal and symbiotic bacteria that dominate the healthy large intestine perform functions that benefit the health of the host. These include the breakdown of dietary fibre to produce short chain fatty acids, which play an important role in uptake processes and in shaping the gut environment and regulating host metabolism. These microbial activities are also crucial in helping to suppress the growth of pathogenic micro-organisms and in suppressing inflammation. In other words, we have co-evolved with our resident gut micro-organisms to the point where we are mutually dependent. Like it or not, we depend on our resident commensal microbiota to maintain a balanced healthy state throughout life.

© Springer Nature Switzerland AG 2020
H. J. Flint, *Why Gut Microbes Matter*, Fascinating Life Sciences,
https://doi.org/10.1007/978-3-030-43246-1_12

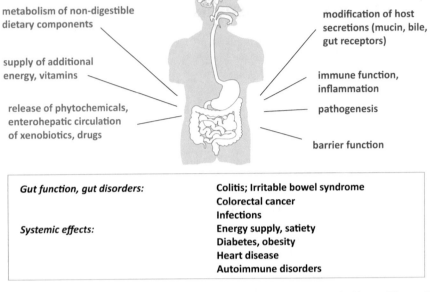

metabolism of non-digestible
dietary components

supply of additional
energy, vitamins

release of phytochemicals,
enterohepatic circulation
of xenobiotics, drugs

modification of host
secretions (mucin, bile,
gut receptors)

immune function,
inflammation

pathogenesis

barrier function

Gut function, gut disorders:	Colitis; Irritable bowel syndrome
	Colorectal cancer
	Infections
Systemic effects:	Energy supply, satiety
	Diabetes, obesity
	Heart disease
	Autoimmune disorders

Fig. 12.1 Impact of the gut microbiota on human nutrition and health (*modified from*—Flint et al. 2011 [1])

On the other hand, by no means all the activities of our commensal (supposedly harmless) gut microorganisms are beneficial. We have seen that some bacteria that are not infectious pathogens nevertheless produce carcinogenic and toxic products, while others have the capacity to promote inflammation or fat absorption. With increasingly detailed knowledge we are working our way towards discriminating between 'good' and 'bad' bacterial profiles among the commensal microbiota. Indeed, it turns out that we may often have to think in terms of 'good' and 'bad' strains, such are the fine differences between strains within species. Metagenomics offers a culture-independent route for detecting harmful genes and gene combinations. But the question remains—can we define a healthy gut microbiota simply by quantifying the relative abundance of species and strains, or genes, present in the microbiome? There is a strong argument that we should be thinking about the overall balance of the community and its activities as a system. The production or removal of individual metabolites is generally a collective property of the microbial community that involves the concerted action of many species. For this purpose, we may need to think about microbial functional groups (MFGs), such as butyrate-producers, more than about individual species. As discussed in Chap. 11, we should be treating the system quantitatively by considering absolute numbers of organisms and metabolite fluxes. System balance, between pro-inflammatory and anti-inflammatory components of the microbiota, is also highly relevant when considering impacts on immune function (Table 12.1).

Table 12.1 What defines a healthy gut microbiome?

Mechanisms	Beneficial species?	Community profile?
Host interactions: Immune function, mucosal barrier function.	Bifidobacteria, lactobacilli, *Faecalibacterium*	SCFA production, Balance of pro-/anti-inflammatory signals
Inhibition of pathogens: competitive binding, antimicrobials, competition for substrates, gut environment	Bifidobacteria, lactobacilli, lactate-utilizing bacteria	SCFA production, lactate utilization. Intestinal pH, Community stability
Nutrient supply: Energy supply to mucosa	Fibre-degrading bacteria Butyrate-producing Firmicutes	SCFA, butyrate production, fibre-degradation, vitamin supply
Cancer-prevention:	Fibre-degrading bacteria Butyrate-producing Firmicutes	butyrate production, phytochemical release, metabolism of bile acids, protein

Finally, we should be clear that some micro-organisms and microbial activities have the potential to be both good and bad, depending on the context and health outcome. To give just one example, *Bacteroides fragiis* is a potential pathogen, but it also produces a polysaccharide that suppresses inflammation and may have therapeutic value [2]. Once again, simplistic assumptions about 'good' and 'bad' micro-organisms can be misleading.

Practical Outcomes

The most important aim, of course, is to arrive at practical solutions for improving human health. Fortunately, this does not depend on gaining a complete understanding of our gut microbiota, any more than the discovery of anaesthetics required a complete understanding of human physiology. This said, it is also true that anaesthesia has benefitted enormously from improved knowledge of physiology. In the case of the gut microbiome we can say that there has been significant progress in several areas that are of practical benefit.

Diagnostic and Predictive Value of the Microbiome We have seen that diet has a significant impact on faecal microbiota profiles, and that some bacterial species are highly specialised in the substrates that they utilize for growth. Such diet-responsive species may indeed prove to be good indicators of dietary intake (e.g. of fibre or resistant starch) and could be helpful in identifying and modifying dietary habits that are associated with poor health outcomes. Their use as nutritional *biomarkers* is however somewhat complicated by the existence of inter-individual variation in microbiota profiles. In other words, with multiple species responding to a given dietary change, not all of them may be present in every person's gut microbiota. For example, stratification into *Prevotella*-predominant and *Bacteroides*-predominant individuals appears necessary when trying to predict responses to *dietary interven-tion* with prebiotics (Chaps. 6, 7). Other criteria for stratification may also be required as we learn more about variation in microbial communities. Ultimately, we can envisage microbiome profile-based nutritional advice as a component of 'personalised nutrition'.

Faecal Microbiota Transfer: Successes and Limitations Faecal microbiota trans-fer (FMT) has proved remarkably successful in treating *Clostridium difficile* infec-tions, with a cure rate of over 90% [3]. The therapy involves introducing 'healthy' faecal material (most commonly from a relative of the patient) into the intestine with the goal of restoring the protective barrier function of the microbiota. Most often the route for delivery is rectal, for example using enemas or via endoscopy. Other methods that have been used include nasoduodenal intubation, which allows deliv-ery into the small intestine, while oral delivery using capsules is also considered a possibility. The FMT approach is in fact much more successful than treating the patients with vancomycin, an antibiotic that inhibits *C. difficile* growth. Why should this be? The main reason seems to be that loss of the normal healthy microbiota in these patients, as a result of the infection, leaves them highly susceptible to re-colonisation by *C. difficile* (a spore-former) even after vancomycin treatment. What FMT does is to restore the resistance conferred by the healthy microbiome to colonisation by *C. difficile*.

The success of FMT in treating *C. difficile* has naturally led to wider interest in its potential for treating a whole range of health problems in which the gut microbiota are thought to play a role. These include type 2 diabetes, obesity and the gut disorders such as IBS and IBD. It should be stressed that these are still early days. Definitive evidence of benefits from FMT in these conditions is currently lacking and it will take some time for the critical evidence to become available [3]. Some caution is also warranted for other reasons. *C. difficile* creates a situation in which the altered gut microbiota profile is the underlying cause of susceptibility to continuing infection. In other conditions such as IBD, however, variations in immune function and host genotype are important contributors to the disease. As a result, attempts to restore a healthy microbiota may either not succeed, or turn out to be by insufficient by themselves as treatments. In other conditions such as obesity and type 2 diabetes, where gut microbiota may play a role, but are clearly not the sole determinants, it would perhaps be over-optimistic to expect the spectacular results from FMT that are seen for *C. difficile* infection.

There are some obvious safety concerns with FMT. Donors of faecal material should be healthy and should be free of infectious viruses including HIV and hepatitis, infectious bacteria and protozoal parasites and of bacteria carrying multiple antibiotic resistances. These are only the most obvious items from a long list of exclusion criteria applied to potential donors of faecal transplants [3].

Next Generation Probiotics and Therapeutics As we have seen, faecal microbiota transfer has features that make it a far-from-perfect solution even where the success rate is high. There is inevitable uncertainty about the reproducibility of what is at best an incompletely defined microbial inoculum. No two faecal samples, even from the same donor, will ever be completely identical and screening is needed to exclude the presence of infectious agents in the donor sample, preferably not too long before the transfer.

If we could just define which beneficial bacteria we wanted to deliver, then surely it would be preferable to deliver a defined, therapeutic cocktail made up from purified, cultured bacteria? This idea has of course occurred to many researchers and commercial companies and is being actively pursued [4]. Existing probiotics come mainly from what are known as the 'lactic acid bacteria' and most are strains of *Lactobacillus* and *Bifidobacterium*. These organisms are relatively tolerant of oxygen, they are easy to grow and they survive well in foodstuffs. Furthermore, based mainly on their long-term usage, they are 'generally regarded as safe'. But they are not necessarily the species most important for maintaining a healthy gut microbiota in adult humans. As we have learned more about the gut microbiome, several of the dominant anaerobic species have been identified as candidates to be therapeutic agents or the next generation of probiotic organisms. These were mentioned in Chap. 5, and include *Faecalibacterium prausnitzii*, *Anaerobutyricum hallii*, *Roseburia* species and *Akkermansia muciniphila*, These bacteria are all oxygen-sensitive, to differing degrees, and require tailored anaerobic culture conditions for the preparation of bulk inocula. They may also require techniques such as encapsulation for optimal oral delivery, although animal experiments have shown that some of these organisms can survive passage through the stomach in sufficient numbers to colonise the large intestine. While the presence of each these species at greater than 10^9 per g in the healthy gut microbiota is somewhat reassuring, safety assessments including screening for genes implicated in pathogenicity or antibiotic resistance will still be required. Spore formation could be helpful in delivery and storage, but it could also be unhelpful for containment of the therapeutic organisms in that inoculated strains will not remain limited to the desired target.

Targeted Prebiotics The initial development of prebiotics had a distinct element of serendipity about it. Bifidobacteria were held to be beneficial to health and their success in the infant gut can be largely attributed to their ability to utilize human milk oligosaccharides (HMOs). HMOs are not easy molecules to purify or synthesise on a large scale, however, so it was fortunate that fructans such as inulin and fructo-oligosaccharides (FOS) turned out to promote bifidobacteria in the adult gut. We now know that many other species within the adult microbiota can utilize inulin and FOS and that these organisms may also be promoted [5]. We also know that the chain length of fructans has a big influence on the species that are promoted, with a more limited range of organisms able to utilize long chain inulin compared with shorter chain fructo-oligosaccharides.

Interestingly, several oligosaccharides other than FOS, such as galacto-oligosaccharides (GOS) and arabino-xylo-oligosaccharides (AXOS) have also been shown to promote bifidobacteria in vivo. The promoted strains must of course be able to exploit these energy sources, but we know that this ability is also shared by many other species that are abundant in the human colon, notably *Bacteroides* species. Indeed, when we run chemostat competition experiments with a single prebiotic such as inulin as the sole energy source it is generally *Bacteroides* species that come to dominate the community [6]. So, how is it that *Bifidobacterium* species so often come to dominate in human dietary trails in vivo? A likely explanation is that the growth of *Bacteroides* species is curtailed, and bifidobacteria compete more successfully, when the pH is slightly acidic. This hypothesis is supported by chemostat experiments that used a range of controlled pH values between 5.5 and 6.8. Even with inulin as the sole energy source, *Bacteroides* species were dominant at pH values between 5.9 and 6.8 and bifidobacteria became numerous only at the lowest pH of 5.5 [5]. The key point here is that active fermentation within the proximal colon can, and does, produce pHs as low as 5.5 because of the build-up of fermentation acids. A pulse of readily fermentable fibre, such as a prebiotic oligosaccharide, will lead to increased acid production and a drop in gut pH. When bifidobacteria are involved in this fermentation the products will include lactate, which is signifcantly more acidic than acetate, propionate or butyrate. This can explain why so many different fermentable fibres appear to promote bifidobacteria in vivo—because they have in common a tendency to decrease the pH in the proximal colon, thus creating conditions that favour bifidobacteria as competitors for the prebiotic.

As our understanding of the human gut microbiota becomes more detailed, it should be possible to target prebiotics more precisely. For example, we may wish to promote other species (such as *Faecalibacterium prausnitzii*) whose health benefits are becoming apparent. To this end, detailed knowledge of patterns of carbohydrate utilization within the microbiota can inform the selection of the most appropriate and selective prebiotic. For example, pectins appear to be utilized by only a small subset of Firmicutes bacteria (although by many Bacteroidetes) in the human colon. One Firmicutes species, *Eubacterium eligens*, however shows excellent capacity to compete for pectins in the mixed community and fortuitously this species also turns out to have anti-inflammatory properties when tested with cultured human cell lines in vitro [6]. In addition to using crude or semi-purified fractions of plant fibre as prebiotics, there is much scope for chemical synthesis and modification of carbohydrates. This should create exciting new possibilities for targeted prebiotics. Some of these are likely to require formal safety approval as *novel foods*.

Looking into the Shadows: Little Known Organisms Most of the bacteria that dominate our intestines have so far received little or no detailed study. This is not so much because they cannot be cultured. More often it is because too few laboratories are equipped for, or interested in, pursuing culture-based work with oxygen-sensitive organisms. For example, research papers on the microbiology (as opposed to molecular analyses reporting occurrence in the gut microbiome) of highly abundant and potentially beneficial Firmicutes species such as *Faecalibacterium prausnitzii*, *Ruminococcus bromii* and *Anaerobutyricum hallii* are still very rare. This is in stark contrast with several oxygen-tolerant bacterial species such as *Escherichia coli* and *Bacillus subtilis* for which publications number in the 10's or 100's of thousands. This presents researchers with a wonderful opportunity for progress in understanding the roles of these organisms in the gut community and their impact on the human host. At the same time this work should lead to the development of novel therapeutics and dietary interventions, as discussed above.

Metagenomic analysis has also revealed OTUs or 'metagenome species' that have no close cultured relatives. These organisms may have been overlooked because of low abundance, or because they fail to grow on standard laboratory media. In some cases, their molecular signatures have been correlated with positive or negative health outcomes, making it very important to learn more about them. Metagenome-derived sequence information can help in devising strategies for isolating such organisms through culturing by suggesting likely growth requirements [7]. It can also provide sequence-based probes that assist in tracking the unknown organisms during their isolation.

Success and Failure of Antibiotics

The discovery of antibiotics during the twentieth century produced the most remarkable advance in the treatment of infectious diseases. Penicillins, which are produced by strains of the fungal mould *Penicillium*, interfere with the formation of the bacterial cell wall. First discovered by Alexander Fleming in 1928, penicillins were put into large scale production in the 1940s and subsequently helped to save countless lives from life-threatening bacterial infections. They were closely followed by bacterially-produced antibiotics such as the tetracyclines, macrolides and aminoglycosides, most of them produced by species of *Streptomyces* (Table 12.2). Some antibiotics such as tetracyclines are active against nearly all types of bacteria

Table 12.2 Major classes of antibiotic used to combat bacterial infections

Antibiotic class	Example	Source[a]	Cellular target
Penicillins (β-lactam)	Penicillin	Fungi (*Penicillium*)	Peptidoglycan (bacterial cell wall)
Cephalosporins (β-lactam)	Cephalosporin	Fungi (*Acremonium*)	Peptidoglycan (bacterial cell wall)
Carbapenems (β-lactam)	Carbapenem	Bacteria (*Streptomyces*)	Peptidoglycan (bacterial cell wall)
Glycopeptides	Vancomycin	Bacteria (Actinomycetes)	Peptidoglycan (bacterial cell wall)
Tetracyclines (polyketide)	Tetracycline	Bacteria (*Streptomyces*)	Protein synthesis (ribosome)
Macrolides (polyketide)	Erythromycin	Bacteria (*Streptomyces*)	Protein synthesis (ribosome)
Aminoglycosides	Streptomycin	Bacteria (*Streptomyces*)	Protein synthesis (ribosome)
Quinolones	Fluoroquinolone	Chemical synthesis	DNA replication
Sulfonamides	Sulfamethoxazole	Chemical synthesis	Synthesis of folic acid

[a]The original microbial source is listed, but there are also synthetic alternatives in most cases

(broad spectrum antibiotics) while others have a narrower spectrum, being effective only against particular groups. Bacteria may be intrinsically resistant to a given antibiotic because they lack the relevant target, for example by virtue of their cell wall structure. Thus, some antibiotics are more effective against Gram-negative pathogens while others are more effective against Gram-positives. Antibiotics are used mainly against bacteria, but specific anti-fungal and anti-protozoal agents have also been developed that provide important defences against eukaryotic infectious agents.

Within a few years of the first clinical applications of antibiotics came the alarming discovery that the targeted infectious bacteria could become resistant to the antibiotic. Research showed that this could happen in several ways. Huge populations of bacteria inevitably contain many small genetic variations (*spontaneous mutations*) such that the occasional mutant cell is resistant to a given antibiotic. The offspring of this resistant cell will continue to grow and replicate despite the presence of the antibiotic and will quickly replace sensitive bacterial cells. Much more alarming than this relatively rare 'mutational resistance', however, was the discovery that genes conferring resistance can be passed readily between different strains and species of bacteria. Furthermore, it was found that resistance to several different classes of antibiotic could be transmitted simultaneously to a previously sensitive bacterial strain. Where did these resistance genes come from? The answer is that they were already present within the microbiome. For one thing, antibiotic-producing organisms need to carry resistance genes to protect themselves against their own antibiotic and such a gene only has to be acquired and expressed by another bacterium for that bacterium to also become resistant [8]. *Horizontal transfer of genes* between different microorganisms is known to occur readily by a variety of mechanisms that play major roles in bacterial adaptation and evolution. These include the uptake of DNA that has been released by another organism (*transformation*) and the exchange of small pieces of DNA via bacterial mating (*conjugation*) or bacteriophage infection (*transduction*). Some DNA within bacterial cells exists as circles called plasmids that replicate themselves independently of the main chromosome and some of these plasmids actually promote conjugation and gene transfer. If an antibiotic resistance gene resides in a plasmid, it can spread very rapidly through a bacterial population, particularly when the relevant antibiotic is present to kill off competing sensitive cells. Single plasmids often carry several resistance genes, meaning that one antibiotic can increase the frequency of unrelated resistance genes. Another type of 'selfish' DNA element known as 'conjugative transposons' that are found to harbour resistance genes reside in the chromosome but can excise and transfer themselves between bacteria. Conjugative plasmids and transposons are capable of transfer into a wide range of host species and have played a very prominent role in the spread of antibiotic resistance.

It is worth considering what happens when you take a course of oral antibiotics to treat an infection. We should pause to note that unless the cause of the infection has been established, the antibiotic may well be ineffective as, for example, when the disease symptoms are caused by a virus. Inevitably, while some of the ingested antibiotic will be absorbed into the bloodstream, part of the antibiotic dose will persist into the small and large intestine where it will act on the resident microbial community. If the antibiotic is broad spectrum, then most members of the gut microbial community will potentially be affected by it. Since most of these organisms are harmless or beneficial, inhibiting their growth is a negative outcome. It is almost certain, however, that some of the bacteria in the complex gut community will carry resistance genes and often these will be situated on transferable elements such as plasmids and transposons. Continued antibiotic consumption will therefore not only select for these resistant organisms but also for the transfer of their genetic elements and resistance genes to sensitive bacteria within the community. Indeed, we have found that an individual taking a broad-spectrum antibiotic over many years can have a gut microbial community that looks perfectly normal and healthy in terms of its species balance. It is just that every single one of these diverse strains has become resistant to the antibiotic in question, with each bacterium typically carrying more than one gene conferring resistance to the antibiotic [9].

In addition to the discovery of new natural producers of antibiotics, chemical synthesis has been widely used to create modified antibiotics or enzyme inhibitors that can overcome resistance. For example, penicillins can be destroyed by microbial enzymes called beta-lactamases, leading to resistance. As part of the arms race against resistance, a whole class of molecules has been developed (exemplified by clavulanic acid) that block the action of microbial beta-lactamases and so restore the efficacy of the antibiotic when administered in conjunction with penicillins. Unfortunately, bacterial adaptation continues to counteract such new strategies, which provide only temporary respite from the advance of resistance. There is a continuing and urgent need for new classes of antimicrobials and new antimicrobial strategies, together with more responsible use of antimicrobials [8].

Mining and Managing the Microbiome

The microbiome represents a vast repository of genes coding for enzymes of potential value for clinical, commercial and environmental applications. These include numerous enzymes concerned with the degradation of carbohydrates of dietary or host origin. There is also potential for discovering new classes of antibiotic, aided by the analysis of microbial genomes [10]. As noted in Chap. 9, some gut microbial genomes have the capacity to synthesise molecules that influence the physiology of host cells, with potential consequences ranging from immune function to fat metabolism and even to brain activity and behaviour. While it is always possible that the producer strains might themselves be developed to make such valuable products, the necessary technical developments for large scale cultivation generally take some time. More often, exploitation of such activities through biotechnology involves the transfer and *heterologous expression* of the relevant genes in more convenient host organisms such as *E. coli* or yeast. The desired product may be derived from a single gene (e.g. a degradative enzyme) or it may be the product of an enzyme pathway that is coded for by a cluster of genes (e.g. a secondary metabolite).

When considering the potential for biotechnological applications, there is of course no reason to limit our searches to members of the gut microbiome. The global microbiome and its vast array of genetic potential is open to us for solving the pressing medical and environmental problems that are confronting us. It is of little consequence, for example, whether a new class of antibiotics comes from soil, the deep ocean or the gut of an insect, provided that it helps to combat infectious disease in humans. If we need to find enzymes that function at extremes of temperature or pH, then we can look for the relevant genes in extreme environments and extremophiles—indeed this is already a standard approach. In these endeavours, the astonishing natural diversity of microbes that has resulted from of billions of years of evolution is very much our ally.

The discipline of Microbiology, i.e. the study of the Microbiome, is surely more important to our survival now than it has ever been before. Fortunately, our new research tools, many of which have emerged often quite unexpectedly from the study of microorganisms, mean that progress is now faster than ever before. The importance of the global microbiome in relation to nutrient cycling and managing the huge problems of man-made climate change and pollution cannot be understated. Just as the gut microbiome must be considered a key factor in human health, so the global microbiome is fundamental to our understanding of global ecology and the man-made future.

References

1. Flint HJ et al (2011) Impact of intestinal microbial communities upon health. In: Rosenberg E, Gophna U (eds) Beneficial microorganisms in multicellular life forms. Springer, Berlin, pp 243–252
2. Troy EB, Kaspar DL (2010) Beneficial effects of *Bacteroides fragilis* polysaccharides on the immune system. Front BioSci Landmarks 15:25–34
3. Konig J et al (2017) Consensus report: faecal microbiota transfer – clinical applications and procedures. Aliment Pharmacol Ther 45:222–239
4. Petrof EO et al (2013) Stool substitute transplant therapy for the eradication of *Clostridium difficile* infection: 'RePOOPulating' the gut. Microbiome 1:3
5. Chung WSF et al (2016) Modulation of the human gut microbiota by dietary fibres occurs at the species level. BMC Biol 14:3
6. Chung WSF et al (2017) Prebiotic potential of pectin and pectic oligosaccharides to promote anti-inflammatory commensal bacteria in the human colon. FEMS Microbiol Ecol 93 art no fix127
7. Pope PB et al (2011) Isolation of Succinovibrionaceae implicated in low methane emissions from Tammar Wallabies. Science 333:646–648
8. Davies J, Davies D (2010) Origins and evolution of antibiotic resistance. Microbiol Mol Biol Rev 74:417–433
9. Kazimierczak KA et al (2008) A new tetracycline efflux gene, *tet*(40), is located in tandem with *tet*(O/32/O) in a human gut Firmicute bacterium and in metagenomic library clones. Antimicrob Agents Chemother 52:4001–4009
10. Kim SG et al (2019) Microbiota-derived lantibiotic restores resistance against vancomycin resistant *Enterococcus*. Nature 572:665–669

Glossary[1]

Acetogenic bacteria Bacteria with the ability to convert hydrogen and carbon dioxide (or formate) into acetate

Adaptive immune system System responsible for learning to recognise and respond to new challenges (infectious organisms and viruses)

Aerobic Able to survive and grow in air

Allochthonous Foreign microorganism that may be present transiently in the gut

Amplicon DNA product resulting from PCR reaction

Amylosome Multienzyme complex on the bacterial cell-surface held together by dockerin-cohesin interactions. Concerned with starch breakdown

Anaerobic Able to survive and grow in the absence of air

Anaerobic respiration Recovery of energy from oxidation that involves electron acceptors such as sulfate and nitrate rather than oxygen

Antibodies Protein molecules (immunoglobulins) with the ability to recognise 'foreign' material (e.g. from invading microorganisms and viruses) and trigger a protective response

Apoptosis Programmed death of cells

ATP Adenosine triphosphate

Autochthonous Belonging to the indigenous gut microbiota of an individual

Autoclaving Sterilisation method employed in hospitals using high temperature and pressure to destroy microbial cells and spores

Autotrophs Organisms able to derive all their energy and growth requirements without relying on other life forms

Bacteriophage therapy Use of viruses that attack particular pathogenic bacteria in order to suppress the target bacteria

Bioactive Referring to any molecule that affects cell physiology, bodily function or behaviour

---

[1]**Please note**—this is not a dictionary. It is not intended to be exhaustive and does not list individual species or molecules. It is simply intended to help clarify certain terms that are used in the text.

© Springer Nature Switzerland AG 2020
H. J. Flint, *Why Gut Microbes Matter*, Fascinating Life Sciences,
https://doi.org/10.1007/978-3-030-43246-1

Biofilm A densely-packed assembly of microbial cells and their products associated with a solid surface

Bioinformatics Computerised techniques for handling and interpreting large amounts of data, especially (although not exclusively) sequence data

Bioluminescence Light emitted by living organisms

Biomarkers Measurements that may be used to indicate states of health

Biomass Living matter

Biosynthesis Manufacture by living cells of organic compounds required for the structure and functioning of an organism

Body mass index (BMI) One's weight (kg) divided by the square of one's height (m^2)

Carcinogenic Promoting cancer

Catalysts Agents (including enzymes) that greatly speed up chemical reactions

CAZymes Carbohydrate active enzymes

CBMs Carbohydrate-binding modules. Amino acid sequences found in many CAZymes that bind specifically to carbohydrate chains

Cellulosome Multienzyme complex on the microbial cell surface held together by dockerin-cohesin interactions. Concerned with the breakdown of plant cell walls

Chemiosmosis Process leading to the formation of ion gradients that drive the production of ATP by membrane ATPases in mitochondria and microbial cells

Chemostat An artificial system for studying microbial communities (or individual microorganisms) with constant inflow and outflow of nutrient medium, in which pH, oxygen and temperature can be precisely controlled

Chemosynthesis Gaining energy from chemical reactions

Chloroplast Organelle within green plants that is responsible for photosynthesis

Cloning Referring here to the use of recombinant DNA techniques to isolate and replicate individual fragments of DNA

Coenzyme An organic, non-protein, molecule that is required for the activity of an enzyme

Cofactor Any non-protein molecule that is required for the activity of an enzyme (includes coenzymes, but also metal ions etc.)

Cohesins Regions (often repeated) within large proteins that bind specifically to dockerins

Commensalism Stable association between gut microbes and their host, but without any major (known) benefit either way

Complement system Complex set of proteins that mediates the destruction of invading cells in mammals

Condensation (chemical reaction) Joining together of two molecules with the loss of water

Conjugation Term used for bacterial matings in which a conjugation tube forms connecting a pair of cells. Bacterial DNA can then pass through the tube from donor to recipient cell

Conjugation (of molecules) Enzyme-catalyzed joining together of two molecules

Consensus sequence Output of DNA sequencing based on computer alignment of data from multiple sequence runs to 'iron out' inevitable errors

Conventional animals Term sometimes used for laboratory animals that carry their native gut microbiota

Coprophagy Ingestion of faecal pellets, allowing a second passage through the gut

Core species Microbial species likely to be found in high numbers in the gut of most healthy individuals

Cryoprotective Helps to maintain integrity of biological material during freezing and thawing

Culture-independent Referring to molecular community analysis that avoids the need to grow microorganisms

Cytochrome A group of proteins containing iron that are concerned with electron transport

Deconjugation Enzymatic separation of two molecules joined by conjugation

Dietary intervention A deliberate modification of the diet

Dilution rate Calculated for a chemostat by dividing the throughput of medium (litres per hour) by the fixed volume of medium in the vessel

Disaccharide Two sugars joined together by a glycosidic linkage (e.g. maltose, sucrose)

Distal colon Another term for the descending (or left-sided) colon

DNA Deoxyribonucleic acid, in which the sugar-phosphate backbone contains the pentose sugar deoxyribose. The genetic material of all cellular life forms and many viruses

DNA polymerase Enzyme responsible for making new copies of DNA

Dockerins Short amino acid sequences within larger proteins (mostly enzymes) that bind specifically to cohesins

Dysbiosis Term referring to deviations from the 'normal' balance of the microbiota

Electron transport chain An ordered series of proteins (including cytochromes) and coenzymes that can exist in oxidised or reduced states, that transports electrons. Present in the inner mitochondrial or bacterial membrane.

Electron transport phosphorylation Formation of ATP arising from proton gradients created via the electron transport chain

Endosymbiont A symbiotic organism that exists inside its partner's cells

Energy harvest Term used by some to describe the recovery of usable energy from the diet

Enterohepatic circulation Cycling of compounds between the gut and liver

Enterotypes Term used by some to describe variations in gut microbiota composition between individuals

Enzyme A protein that greatly speeds up (i.e. catalyses) a chemical reaction

Epithelial cells/epithelium Surface layer of cells in tissues of multicellular organisms

Extremophiles Microorganisms whose preferred habitat is extreme (e.g in pH, salinity, temperature)

Facultative anaerobe A microorganism that is able to grow both in the presence and absence of oxygen

Fermentation Anaerobic release of usable energy from complex molecules (e.g sugars) without any net change in the state of oxidation or reduction

Fibre Referring to dietary material (mostly carbohydrate) that is not digested in the upper gut (but see text for nutritional definition)

Fitness Ability to survive and pass genes on to the next generation

Flagella 'Whip-like' mobile structure extending from some microbial cells that helps them to move

Free energy change A term from thermodynamics. The standard free energy change of a chemical reaction (at a given temperature) tells us how readily the reaction will occur.

Genetic distance Used to construct phylogenetic trees. The difference (distance) between two sequences is expressed in nucleotide or amino acid substitutions (with some corrections, e.g. for multiple substitutions at the same site)

Genetic dominance The phenomenon where, in a diploid organism (with two sets of chromosomes and genes), one 'normal' copy of a gene has the same effect as two normal copies

Genetic polymorphism The existence in a population of alternative versions (alleles) of a given gene

Genome (genomics) The complete set of DNA sequences that is passed on to progeny cells during cell division, and from parent to offspring.

Genotype The genetic complement of an individual, as distinct from the consequences of expression of those genes *(phenotype).*

Germ-free animal A laboratory animal with a sterile (germ-free) intestine, usually resulting from caesarean delivery in an isolator

Gluconeogenesis New synthesis of glucose from non-carbohydrate sources (e.g. amino acids)

Glycolysis Conversion of glucose to two molecules of pyruvic acid via the 'Embden Meyerhof' biochemical pathway. Can occur without oxygen

Glycoside hydrolases Enzymes that break down carbohydrate chains by hydrolysing glycosidic linkages

Glycosidic linkage Bond involving an oxygen joining two sugars together

Gnotobiotic animal Meaning 'known biota'. The result of reintroducing single microbial strains or known mixtures of strains into the gut of a germ-free animal

Growth promoters Term used in farm animal production for feed additives that increase live weight gain per unit feed input

Gut barrier function Maintenance of the integrity of the gut lining, thus preventing microbial access

Gut transit Passage of contents through the gut. Can be measured using inert markers as 'mean transit time' for the whole gut, or for regions of the gut

Heterologous gene expression Formation of a gene product after transfer of the gene to a different host species (e.g. the bacterium *E. coli*)

Heterotrophs Organisms that depend on other living organisms for their nutrition and energy

Heterozygote Individual whose two sets of chromosomes carry different versions (alleles) of a given gene

Homozygote Individual whose two sets of chromosomes carry identical versions (alleles) of a given gene

HMOs Human milk oligosaccharides, secreted in maternal breast milk

Horizontal gene transfer Transfer of genes between different species (as opposed to the normal 'vertical' gene transfer to daughter cells of the same species)

Hormones Secreted molecules that exert wide-ranging control over body physiology by binding to receptors in target tissues (e.g. insulin, adrenaline). Most are peptides or steroids.

'Humanised' mouse The result of introducing human gut microorganisms (usually via a faecal inoculum) into a germ-free mouse

Hydrogen bonding Relatively weak attractions between chemical groupings resulting from opposite electrical charges. Has huge biological significance (e.g. for nucleic acid and protein structures)

Hydrogen sink A metabolic route for disposing of hydrogen by producing a relatively reduced compound

Hydrolysis Cleavage of a chemical bond with the addition of water

Hypha Tube-like growth form (containing multiple nuclei) of most fungi

Inflammation A complex set of responses by the host to injury, infectious agents or immune challenges

Innate immune system System that recognises general features of potentially infectious organisms and viruses, triggering an initial defensive response.

In vitro Literally 'in glass', referring to experimental work conducted without involving living organisms

In vivo Literally 'in life', referring to a living body or natural microbial community

Inoculum Introduction of live microbial cells

Ion An atom or molecule that carries a net positive (cation) or negative (anion) charge as a result of losing or gaining an electron

Keystone species Species whose absence from a community has wide-ranging consequences for the whole community

Lactic acidosis Life-threatening condition in which excessive lactic acid production drives down the pH of the gut

Lignocellulose Plant cell material composed of structural carbohydrates (cellulose, hemicellulose, pectin) linked to the non-carbohydrate (phenolic) polymer lignin

Lymphocytes White blood cells mainly responsible for antibody production

Lysozyme Enzyme that attacks specific glycosidic linkages in bacterial cell walls

Melting Used to refer to the separation of DNA strands when hydrogen-bonding is broken down (e.g. by raised temperature)

Methanogens Archaea capable of making methane from hydrogen and carbon dioxide or formate (or in some cases methanol)

Metabolites Any small molecule to be found within the cell or body

Metabolome (metabolomics) The collection of small molecules that is present in a biological sample. [n.b. The term *metabonomics* is used by some when referring to quantitative measurement]

Metagenome Referring to the collective genomes present in an environmental sample

Microarray Analyses in which large numbers of samples are arranged in a grid pattern

Microbiota composition (or microbiota profile) Here referring to the proportions of different microbial (generally bacterial) groups within a gut or faecal sample as estimated by molecular methods

Mitochondrion Organelle within eukaryotic cells responsible for respiration

Monosaccharide Single sugar molecule (e.g. glucose, fructose)

mRNA RNAs that code for proteins, produced by transcription of DNA molecules

Mucin Complex molecules in mucus consisting of protein and carbohydrates that play a protective and lubricating role on external tissue surfaces, including the gut

Mycelium Network of fungal hyphae

Mycorrhizae Fungi associated with the root systems of land plants

Neurotoxin Toxin damaging to the nervous system

Niche In the ecological sense, the particular conditions that allow a given organism to compete and survive within the ecosystem

Non-ribosomal protein synthesis Referring to enzyme systems that manufacture specific peptides without the involvement of ribosomes

Nucleus Location of the chromosomal material within the cell

Nucleoside Term used for a base joined to ribose or deoxyribose (e.g. ATP is described as a nucleoside triphosphate)

Nucleotides Building blocks of nucleic acids, comprising a base joined to ribose or deoxyribose linked to a phosphate

Obligate anaerobes Organisms that are unable to grow in the presence of oxygen

Oligosaccharide Chain of three or more sugars (but small enough to be water-soluble)

Opportunistic pathogens Organisms that can exist as harmless commensals, but can also cause serious infections when host defences are compromised

Organelle Specialised structure within a cell (e.g. mitochondrion)

Oxidation (oxidised) Variously defined as net loss of electrons, gain of oxygen or loss of hydrogen

Pan-genome Representing variation and conservation of genes between genomes across multiple strains of a species

Pathogen A microorganism that causes disease

PCR Polymerase chain reaction, used to amplify DNA sequences

Periplasm Region lying between the two cellular membranes around a Gram-negative bacterial cell

Peristalsis Contractile movements of the gut wall that mix and propel gut contents

Phagocyte Cell capable of engulfing and destroying microorganisms or other cells

Photosynthesis Trapping of biologically usable energy from sunlight

Phylogenetic tree Representation of the evolutionary divergence between different organisms, as inferred from sequence information

Phylum The highest level of taxonomic grouping within a domain

Phytochemicals Diverse chemicals exclusively or mainly of plant origin

pKa pH at which a weak acid (e.g. acetic acid) is 50% ionised (as acetate anions)

Plasmid A small circle of DNA capable of replicating itself independently of the main chromosome. Some plasmids promote their own transfer to other cells

Polysaccharide Long sugar chain (10–100's of sugars). Generally insoluble or only partially soluble in water

Polysaccharide lyases Enzymes that break down carbohydrate chains without hydrolysis

Prebiotics Non-digestible carbohydrates thought to confer health benefits by promoting the growth and/or activity of selected species of microorganism in the gut

Probes Referring here to short nucleotide sequences that recognize specific targets, usually tagged with fluorescent groups for convenient detection

Probiotics Defined microbial preparations that are thought to confer health benefits

Prophage A viral genome that has inserted itself into the host chromosome

Proteome (proteomics, meta-proteomics) Collection of all proteins present in a given cell population.

Proximal colon Another term for the ascending (or right-sided) colon

PUL Polysaccharide utilization locus. Refers to a cluster of bacterial genes whose products are responsible for the utilization of a complex carbohydrate.

Purine Nitrogenous bases found in nucleic acids (adenine and guanine)

Pyrimidine Nitrogenous bases found in nucleic acids (thymine, cytosine, uracil)

Quorum sensing Mechanisms by which microbial populations monitor and then react to cell density

Receptors Cell-associated structures involved in specific binding (including to hormones, small molecules, other cells) generally resulting in a cellular response

Recombinant DNA techniques Techniques that allow the cutting and re-joining of DNA molecules of different origin in the test tube and their propagation in convenient host microorganisms (most often *E. coli*)

Reduction (reduced) Variously defined as net gain of electrons, gain of hydrogen or loss of oxygen

Resistant starch That fraction of dietary starch that is not digested by the host's digestive enzymes

Respiration Release of usable biochemical energy through oxidation of complex molecules (e.g. sugars)

Ribosomes Particles (composed of proteins and RNA) that mediate the translation of mRNAs into protein sequences

RNA Ribonucleic acid, in which the sugar-phosphate backbone contains the pentose sugar ribose. Present in ribosomes, transfer and messenger RNAs as well as some viral genomes

RNA polymerase Enzyme responsible for making RNA transcripts from DNA

Satiety Suppression of hunger

Secondary metabolites Metabolites that are not essential for cell growth under most circumstances, but may play other roles (e.g. in signalling, adhesion, antimicrobial activity)

Sequence homology Degree of identity between two sequences, assumed to be due to common ancestry

Sequestration mechanism Describing cellular organisation that prevents breakdown products from becoming available to competing microorganisms

Spontaneous mutation Rare changes in DNA sequence resulting from errors in DNA replication and damage repair

Steady state conditions Here referring to a state in which metabolite concentrations, cell concentrations and microbial community composition are all stable with time

Stereoisomers Alternative 'mirror image' arrangements of molecules that otherwise have identical elemental composition and branching structures. Crystals of the two isomers often rotate light differently and are referred to as L- (levorotatory) and D- (dextrorotatory).

Stoichiometry Referring to the precise balance between substrates and products in a chemical reaction

Stratification Here referring to the subdivision of subjects in a study into one or more distinct groupings for the purpose of data analysis

Substrate level phosphorylation Formation of ATP (or sometimes GTP) driven by transfer of high energy phosphate from another compound (e.g acetyl-phosphate)

Symbiosis, symbiotic Living together for mutual benefit

Syntrophy Interaction of two microorganisms resulting in a chemical conversion that would not be achieved by either organism on its own

Taxon (taxa) Any unit of classification for living organisms (e.g. species, phylum)

Telemetry Technique using an electronic device that can be swallowed and then tracked through the gut, allowing continuous monitoring of pH and pressure

Therapeutic Of use in the treatment of disease

Transcription Referring to the copying of nucleotide sequences in DNA to RNA (by RNA polymerase).

Transcriptome (transcriptomics, meta-transcriptomics) Collection of all RNA molecules produced by transcription in a given cell population (often limited to mRNAs to avoid the more abundant ribosomal RNAs).

Transduction Refers to the transfer of chromosomal DNA fragments from one cell to another via bacteriophage particles

Transformation Acquisition by a microbial cell of DNA from its immediate environment. This happens naturally in some bacteria but can also be induced in the laboratory.

Translation Referring to the conversion of nucleotide sequences into protein sequences on ribosomes

Transposon A DNA region that may be found in the chromosome or in a plasmid that promotes its own transfer to different locations or different cells

Toxin Harmful compound, including microbial toxins produced by bacteria and fungi

Vacuole A fluid-filled space within a cell, surrounded by a membrane

Villus (villi) Protuberances on the intestinal wall that play a key role in absorption of nutrients, especially in the small intestine

Virulence genes Genes whose activity contributes to the ability of a pathogen to cause disease

Xenobiotics 'Foreign' compounds of biological origin

Index

© Springer Nature Switzerland AG 2020
H. J. Flint, *Why Gut Microbes Matter*, Fascinating Life Sciences,
https://doi.org/10.1007/978-3-030-43246-1

Printed in the United States
by Baker & Taylor Publisher Services